T0305529

Failure Analysis of Composite Materials with Manufacturing Defects

In contrast to metals, a composite material acquires an internal structure where the imprint of its manufacturing process history is a significant part of the internal structure's makeup and in many cases determines how the material responds to external impulses. The performance for which a composite material is designed must therefore be assessed with due consideration to the manufacturing-induced features in the material volume. Failure theories based on homogenized composites cannot deliver reliable methodologies for performance assessment. This book details approaches that depart from traditional treatments by accounting for manufacturing defects in composite materials. It discusses how manufacturing defects are produced and how they affect the performance of composite materials.

- Serves as the only book to bring knowledge on manufacturing and failure modeling together in a coherent manner.
- Guides readers on mechanisms-based modeling with a focus on defects.
- Treats statistical simulation of microstructure with defects aimed at physical modeling.
- Covers manufacturing methods for polymer matrix composites.
- Describes failure modes in unidirectional composites and laminates in the presence of defects.
- Discusses fatigue damage in the presence of defects.

This book is aimed at researchers in industry and academia in aerospace engineering, mechanical engineering, and materials science and engineering. It also serves as a reference for students taking advanced courses in composite materials.

APPLIED AND COMPUTATIONAL MECHANICS

A Series of Textbooks and Reference Books

Founding Editor

J.N. Reddy

Continuum Mechanics for Engineers, Fourth Edition

G. Thomas Mase, Ronald E. Smelser & Jenn Stroud Rossmann

Dynamics in Engineering Practice, Eleventh Edition

Dara W. Childs, Andrew P. Conkey

Advanced Mechanics of Continua

Karan S. Surana

Physical Components of Tensors

Wolf Altman, Antonio Marmo De Oliveira

Continuum Mechanics for Engineers, Third Edition

G. Thomas Mase, Ronald E. Smelser & Jenn Stroud Rossmann

Classical Continuum Mechanics, Second Edition

Karan S. Surana

Computational Methods in Engineering: Finite Difference, Finite Volume, Finite Element, and Dual Mesh Control Domain Methods

J.N. Reddy

For more information about this series, please visit: https://www.crcpress.com/Applied-and-Computational-Mechanics/book-series/CRCAPPCOMMEC

Failure Analysis of Composite Materials with Manufacturing Defects

Ramesh Talreja

CRC Press
Taylor & Francis Group
Boca Raton London New York

CRC Press is an imprint of the
Taylor & Francis Group, an **informa** business

Designed cover image: shutterstock

First edition published 2024
by CRC Press
6000 Broken Sound Parkway NW, Suite 300, Boca Raton, FL 33487-2742

and by CRC Press
4 Park Square, Milton Park, Abingdon, Oxon, OX14 4RN

CRC Press is an imprint of Taylor & Francis Group, LLC

© 2024 Ramesh Talreja

ISBN: 978-1-032-12686-9 (hbk)
ISBN: 978-1-032-12688-3 (pbk)
ISBN: 978-1-003-22573-7 (ebk)

DOI: 10.1201/9781003225737

Typeset in Times
by Deanta Global Publishing Services, Chennai, India

Contents

Author Biography

Ramesh Talreja is a Tenneco Professor in the Department of Aerospace Engineering and in the Department of Materials Science and Engineering at Texas A&M University, Texas. From 1991 to 2001, he was a Professor of Aerospace Engineering at Georgia Institute of Technology, Georgia. His research is in composite materials, which he began at the Technical University of Denmark, Kongens Lyngby, Denmark, where he earned his PhD in Solid Mechanics in 1974 and was endowed with a Doctor of Technical Sciences degree in 1985 on his collected works on fatigue and damage mechanics of composites. His recent work has focused on the effects of manufacturing defects on the performance of advanced composites.

Prof. Talreja has published extensively in the composite materials field. He is the recipient of the 2013 ICCM (International Committee on Composite Materials) Scala Award and a World Fellow and Life Member of ICCM. The American Society for Composites selected him for the 2017 Outstanding Researcher Award.

1 A Manufacturing Sensitive Design Strategy

Composite materials have enabled a transformation of the structural landscape in the aerospace, automotive, and energy fields, and continue to do so in these and other fields, mainly because of advances in manufacturing technologies. In the early years of composite materials applications in aircraft, the technology of producing pre-impregnated tapes of unidirectional straight fibers in an epoxy matrix, called prepregs, enabled lightweight structural designs using laminates of multidirectional stacked-up plies. Today, a wide range of fiber architectures and polymer processing methodologies are possible that allow applications in the transportation and energy sectors of industry, and the prospects are continuing to grow. All this has taken a great deal of innovation in manufacturing to meet the demands of cost and feasibility. Accompanying these advances is the aspect that each manufacturing method produces defects of different types and intensities depending on the level of control of the processes involved and the cost constraints. The challenge is to design the structures to perform safely in the presence of those defects. Arbitrarily increasing the factors of safety will introduce unreliability and take away some of the advantages composite materials offer. The traditional way of designing structures with given material properties that are defined on homogenized internal structures is no longer adequate. A new paradigm is needed.

In Figure 1.1 such a paradigm is presented. The starting point in the manufacturing sensitive design paradigm is the manufacturing method employed, e.g., autoclave-based curing of a part prepared with prepreg tape-laying process, a part made with resin transfer molding (RTM), or a modified version of it such as vacuum-assisted RTM (VARTM), or any of the many resin infusion processes. The complete manufacturing of a part before entering it in service can also involve tooling (machining of holes, cutouts, etc.), assembly by co-curing and adhesive bonding, etc. Thus, the internal structure of a composite part may have a spatial distribution of geometrical features that are a result of all such operations. Theoretically, this distribution and the geometrical variation of the features should be possible to predict by modeling and simulation given the process history. However, while progress is being made in this direction, the goal is not yet within sight.

Until the internal structure of a composite can be predicted, the alternative available, at least partly, is to observe it with techniques of different resolution such as optical microscopy and X-ray computed tomography. The next step will then be to relate the observed features to the material response characteristics, e.g., for deformational (stress–strain) response, commonly called "stiffness", or for failure response, commonly called "strength", or for resistance to the growth of a crack, commonly called "fracture toughness". In any case, a more general concept than that of traditional

DOI: 10.1201/9781003225737-1

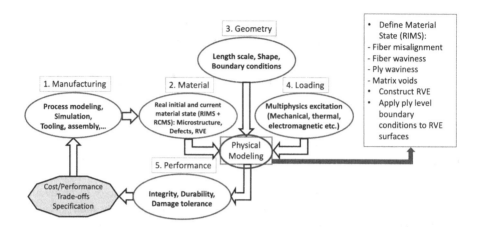

FIGURE 1.1   A flow chart for manufacturing sensitive design of composite structures.

material properties defined on homogenized internal structure is needed [1]. We propose here the concept of a "material state" to be described by statistical descriptors that carry information on the internal structural elements of relevance to the desired material response. These elements can be, for instance, the (homogenized) properties of the constituents, their spatial distributions, and any manufacturing-induced entities such as voids. Since some or all these elements can change under service, the material state descriptors must reflect those changes. Therefore, we introduce the terminology of real initial material state (RIMS) and real current material state (RCMS), which is the evolved material state induced by irreversible changes. The subsequent chapters of the book will deal with statistical simulation schemes related to RIMS and models to describe RCMS.

Continuing with the flow chart in Figure 1.1, the material state descriptors along with the information on geometry at the length scale of interest for failure analysis, and the imposed loading, enter in the physical modeling of the local failure. For illustration, the type of information entered in the failure analysis of a UD composite is shown in the side box to the right in the figure. The output of the physical modeling is a performance assessment related to structural integrity, durability, damage tolerance, etc. of the structural part. The performance indices can be used in conducting a cost/performance trade-off, which can guide controlling or modifying the manufacturing process to achieve a more cost-effective design. The goal is to produce optimal structural designs by minimizing cost.

Reaching the goal of optimal structures necessitates that the manufacturing processes be modified to produce the internal (microstructural) configurations of heterogeneities and defects that will maximize performance at the minimum cost. Achieving the needed manufacturing process modification requires that the manufacturing processes, including assembly interventions, can be modeled reliably. While progress is being made in this direction, as noted above, it is realistic at this time to observe the internal composite structure and simulate it by stochastic methods to create a representation of its features relevant to the initiation and progression of failure. This will be the subject of Chapters 3 and 4. Before those chapters, Chapter 2 will provide a review of the various manufacturing processes for polymer matrix composites (PMCs). Some illustrative modeling efforts will also be described in that chapter. The subsequent chapters will describe observations of failure in composite materials in the presence of defects under static loading in UD composites (Chapter 5) and in laminates (Chapter 6) and under fatigue loading in both (Chapter 7). Physical modeling to account for the manufacturing defects for the purpose of assessing the structural performance will be addressed in Chapter 8 for static loading and in Chapter 9 for fatigue loading.

REFERENCE

1. Talreja, R. and Waas, A.M., 2022. Concepts and definitions related to mechanical behavior of fiber reinforced composite materials. *Composites Science and Technology*, *217*, p. 109081.

2 Manufacturing Methods for Polymer Matrix Composites (PMCs)

2.1 INTRODUCTION

While the synergistic nature of composite materials offers many advantages over monolithic materials, polymer matrix composites (PMCs) are the most used composites due to the low mass density of polymers and the consequent high specific properties such as stiffness and strength per unit of weight when measured in principal (reinforcement) directions. Driven by the requirements of PMCs for aerospace and a wide range of non-aerospace applications, polymers have been developed to contribute to the structural properties of composites as well as to allow the manufacturing of structural geometries within cost constraints. Thus, from the early applications in boats and pipes, where polyester was the common polymer used in a wet layup with E-glass fibers, today advanced thermoplastics are used in additive manufacturing processes. With each combination of fibers, fiber architecture, process type, and process parameter history (e.g., time variation of temperature and pressure), the composite material resulting at the end of the process is a "product" of that combination. The internal composition of that material dictates the structural performance under service. This book is concerned with the role of manufacturing-induced defects on that performance. For that purpose, it is necessary to understand the nature of the defects and how they result from the manufacturing processes. With that in mind, this chapter will review the common manufacturing methods and the type of defects they produce. For more extensive reviews of PMC manufacturing, see [1–4].

2.2 PRACTICAL PMC MANUFACTURING VS MANUFACTURING SCIENCE

There is a distinction between manufacturing science and practical manufacturing. This distinction is important from the perspective of manufacturing defects. In manufacturing science, the objective is to clarify, and possibly predict, the formation of deviations from the intended composite internal structure as a function of the process parameters. In simulating the manufacturing process using the processing parameters it is difficult to account for human intervention which is still present to a significant extent in practical PMC manufacturing. To what extent this intervention is

DOI: 10.1201/9781003225737-2

necessary depends on the type of manufacturing and its cost constraints. Automated manufacturing is an increasing trend and in this type of process it is possible to quantify the process parameters and control them to some extent. This then allows conducting realistic modeling and simulation studies.

In the following, the common PMC manufacturing processes will be discussed in broad categories of resin infusion-based and prepreg-based, and in each case, the focus will be on the defects that result. The hand layup process will not be discussed because of the level of uncertainty involved that does not allow an assessment of defects. For the other two processes, reference will be made to manufacturing science studies that provide insight into the nature of defects, how they result, and their quantification to the extent possible. Since most manufacturing processes are based on thermosetting polymers, the description below will be focused on these. Thermoplastics are increasingly gaining ground in applications, and a section below will be devoted to PMC manufacturing with these polymers. Finally, the recent field of additive manufacturing with PMCs will be discussed from the perspective of defect formation.

2.3 LIQUID COMPOSITE MOLDING

Liquid composite molding (LCM) is a composite manufacturing technology that encompasses molding processes described as resin transfer molding (RTM) and vacuum-assisted RTM (VARTM), and when flexible membranes or shells are used, the molding processes are described as vacuum infusion or other resin infusion techniques under flexible tooling (RIFT) variants [5]. There is no universal consensus on the use of terminology, which leads to some confusion and is addressed in a review [6]. Basically, the idea of LCM is to place dry fibers in a selected configuration, called preform, in a sealed mold cavity and let the liquid resin flow into the cavity filling the porous preform. The mold-filling process can be assisted by vacuum, or the resin can be injected into the mold with positive pressure. After filling the cavity, the resin is cured, followed by removal of the part.

The schematic of the RTM process is depicted in Figure 2.1 showing the steps involved in the process [7].

The resin in the RTM process is cured in time by thermal activation, but in another version of the process, called structural reaction injection molding (SRIM), two or more reactive prepolymers are mixed before infusion and cause rapid solidification by reaction.

The basic steps in the impregnation of dry fibers placed in a selected configuration or as a preform are the same in all processes. When a vacuum is used to assist the resin flow, a vacuum bag and sealant tape are used. A release agent is used to separate the cured part from the mold (or tool) and a peel ply is used if needed between the preform and any outer material not intended to be in the finished product. Depending on whether one chooses to describe the resin flow as "transfer" or "infusion" vacuum-assisted molding is described as VARTM or VARIM, with "I" describing infusion, but often these descriptors are used interchangeably. Figure 2.2 shows details of a VARIM process for illustration [8].

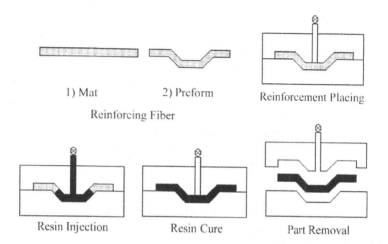

FIGURE 2.1 A schematic of the resin transfer molding process. Source: Shojaei, A., Ghaffarian, S.R. and Karimian, S.M.H., 2003. Modeling and simulation approaches in the resin transfer molding process: a review. *Polymer Composites*, *24*(4), 525–544.

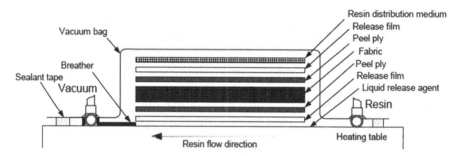

FIGURE 2.2 A schematic of the vacuum-assisted resin infusion molding (VARIM) process. Source: Goren, A. and Atas, C., 2008. Manufacturing of polymer matrix composites using vacuum assisted resin infusion molding. *Archives of Materials Science and Engineering*, *34*(2), 117–120.

The VARIM processes have found wide applications in composite structures, resulting in many variations to suit the applications at hand. An effort was made to characterize some common variations by process parameters and perform a comparison of the cured composite [9]. The parameters that were monitored included infusion time, laminate pressure and thickness, resin mass in the reservoir at the inlet, and void formation. Significant differences were found in fiber volume fraction variations among the processes and void formation was found to be influenced by the resin pressure within the laminate during the post-filling stage. Such studies are useful in understanding the influence of the process parameters on the quality of the composite part but cannot yet predict the internal structural composition of the composite as they do not capture all parameters involved in the practical implementation of a VARIM process.

One of the factors influencing the internal composition of a composite part made with a resin infusion process is the mold filling. The challenges in modeling and simulation of the mold-filling process have been addressed by many researchers and a review of those efforts can be found in [10]. One of the challenges is illustrated by an analytical model of the mold-filling process where the resin flow through the fiber preform is modeled as fluid flow through a porous deformable medium [11]. By modifying this model to include fiber preform compaction [12] it was found that the fiber volume fraction profile was significantly affected by this factor. While such analytical models are useful in parametric studies such as compaction pressure, numerical simulation is necessary to address part geometries. Even then, only simple part geometries can be used to assess the accuracy of simulations. Also, industrial conditions that result in variations in resin viscosity, initial preform thickness, and thermal environment are difficult to incorporate in the simulations [12]. For resin infusion in textile composites, additional considerations of draping of the textile must be made [13].

2.3.1 ANALYSIS OF DEFECT FORMATION

For the purpose of this book, it is of interest to know what type of defects result from the resin infusion processes. As noted above, predicting the internal structure of a composite part made by a resin infusion process is not yet possible. The modeling and simulation studies can at best provide insight into the influence of the process parameters on the quality of the manufactured part. An assessment of the quality is often made by checking certain properties such as the stiffness and strength of the part. Such short-term properties do not always reflect the performance characteristics indicative of the durability and damage tolerance of the part. For that, one needs to have detailed knowledge concerning manufacturing defects that result from the resin infusion processes. While the effects of defects on failure initiation and progression will be discussed in subsequent chapters, here we shall describe how defects are formed during the resin infusion. A comprehensive treatment of defects in resin infusion processes is found in [14].

2.3.1.1 Formation of Voids

An extensive review of the formation of voids in PMCs can be found in [15]. Voids resulting from a resin infusion process can broadly be categorized as unfilled (dry) volumes (or spots) and volumes filled with air or other gasses. The formation of the two types of voids is due to different causes and their size and shape also differ. The dry spot type of void results from the resin not being able to fill all empty spaces within the preform. This can be caused by premature resin crosslinking which affects the resin flow by a local non-uniformity of viscosity. Another cause of the dry spot formation is multiple flow fronts merging and trapping air between them. The entrapped air then prevents the full merger of the flow fronts resulting in unfilled volumes. Yet another cause of dry spot formation is irregular permeability of the preform caused by non-uniformity of the preform geometry or architecture. Similarly, the mold geometry changes can result in the resin flow front not being able to negotiate the mold corners and curvatures.

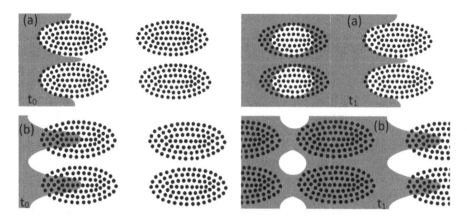

FIGURE 2.3 Void formation by mechanical air entrapment in the resin infusion process when the resin flow front is led by (a) viscosity of the resin and (b) by capillary action. t_1 and t_2 are the initial and final thicknesses of a part. Source: Hamidi, Y.K. and Altan, M.C., 2017. Process-induced defects in liquid molding processes of composites. *International Polymer Processing*, *32*(5), 527–544.

Voids filled with air or other gaseous matter are formed by various processes that include mechanical air entrapment during resin infusion, volatilization of dissolved gasses or moisture, partial evaporation of mold release agents, and any air initially present in the resin. Of all these causes, mechanical air entrapment is believed to be the most common cause. A schematic of this void formation process is shown in Figure 2.3 where the two main scenarios of viscosity-driven and capillary-driven flow fronts are illustrated. In the case of viscosity-driven flow fronts, the resin can flow past the fiber bundles (tows) without completely filling the empty spaces within the bundles, resulting in air entrapment within the bundles. In the scenario where the capillary forces are leading the resin flow front, the empty spaces within the fiber bundles are filled but the resin flow front lags in the inter-tow regions, causing air entrapment there. Voids formed by air entrapment in the two scenarios are likely of different sizes and are described as micro voids (or microscale pores) and meso voids (or mesoscale pores), as indicated in Figure 2.4. Figure 2.5 shows optical images of typical voids formed in a PMC part made by resin infusion. More detailed images of the voids can be seen by using X-ray computed tomography (XCT) using contrast enhancement (Figure 2.6) [15]. As seen, the scale and distribution of the voids has significant variation.

2.3.1.2 Formation of Fiber-Related Defects

The deviations from the desired (ideal) internal configurations of fibers in a resin-infused PMC can come from sources at the pre-infusion stage and during the resin infusion process. The formation of defects in parts made with woven preforms has been discussed in [16]. These defects can arise from how the resin infusion process has been designed, i.e., the tool geometry, the choice of the woven fabrics and how the fabrics are placed, and from the resin infusion process itself. While design improvements can reduce the fiber configuration defects, the infusion process-induced

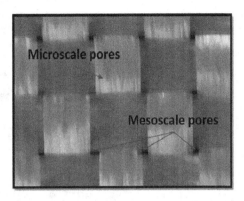

FIGURE 2.4 Voids (pores) of different scales in a woven fabric composite. Source: Advani, S.G., Role of Process Models in Composites Manufacturing, 2018. Comprehensive Composite Materials II, Vol. 2, 24–41.

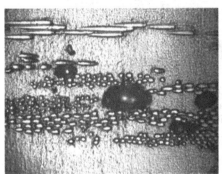 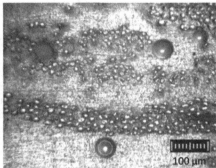

FIGURE 2.5 Representative microscopic images of inter-tow voids and intra-tow voids formed in resin transfer molded PMC parts. Source: Hamidi, Y.K. and Altan, M.C., 2017. Process-induced defects in liquid molding processes of composites. *International Polymer Processing*, 32(5), 527–544.

defects are formed by complex mechanisms and must be understood. These defects are mainly in-plane tow waviness and through-thickness ply wrinkles.

The in-plane tow waviness can result from axial compression in the tow fibers during consolidation and the constraint to tow movement induced by the neighboring ply. The resulting buckling of the tows leads to in-plane waviness. Similarly, the compressive forces transverse to the tows can cause out-of-plane rotation of the tows resulting in through-thickness ply wrinkles.

Figure 2.7 [16] shows images of in-plane tow waviness in a 0° unidirectional (UD) and in a 0/90 satin weave in a 0° bias direction The through-thickness wrinkles are illustrated in Figure 2.8 [16].

While efforts are made to accurately place the preforms in the mold and the careful handling of the preforms to avoid damage to the fibers, the level of accuracy depends on the training of the technicians conducting the resin infusion process and

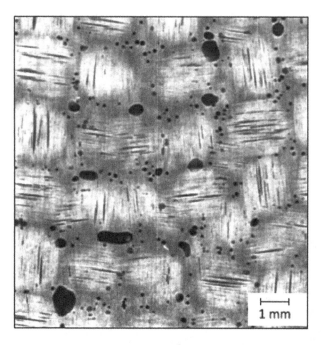

FIGURE 2.6 CT scan image of a carbon/epoxy woven fabric composite. Source: Mehdikhani, M., Gorbatikh, L., Verpoest, I. and Lomov, S.V., 2019. Voids in fiber-reinforced polymer composites: A review on their formation, characteristics, and effects on mechanical performance. *Journal of Composite Materials*, *53*(12), 1579–1669.

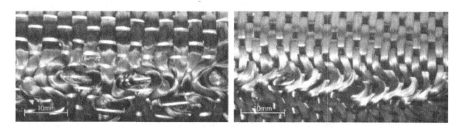

FIGURE 2.7 In-plane tow waviness in a 0° UD (left) and in a 0/90 satin weave in the 0° bias direction. Source: Lightfoot, J.S., Wisnom, M.R. and Potter, K., 2013. Defects in woven preforms: Formation mechanisms and the effects of laminate design and layup protocol. *Composites Part A: Applied Science and Manufacturing*, *51*, 99–107.

the size and shape of the part to be molded. The cured part is therefore likely to have fibers displaced from their intended positions as well as clustered and wavy. Such fiber distortions and non-uniformity in their distribution within the composite volume are difficult to predict even when the process parameters are identified and controlled. These fiber-related defects can properly be treated by stochastic simulation methods and their effects on the composite performance be analyzed by appropriate failure criteria. This will be the subject of a later chapter in the book.

FIGURE 2.8 Ply wrinkling in a 90° weave where the reduced thickness of the tows indicates the presence of wrinkles. The binder yarns in yellow are misaligned due to the wrinkling. Source: Lightfoot, J.S., Wisnom, M.R. and Potter, K., 2013. Defects in woven preforms: Formation mechanisms and the effects of laminate design and layup protocol. *Composites Part A: Applied Science and Manufacturing, 51*, 99–107.

2.3.1.3   Formation of Interfacial Defects

One source of defects in the fiber/matrix interfaces in resin infusion processes is the incomplete filling of the spaces between fibers in a preform. This leaves not only voids, as discussed above, but also unwetted fiber surfaces that act as stress concentration sites and can become cracks under stress. Inadequate wetting of fiber surfaces can also cause insufficient bonding between fibers and the polymer resulting in the lack of ability of the fibers to provide the properties and performance expected of the composite. Unfortunately, it is not easy to detect imperfect fiber/matrix interfaces within a composite due to their size. The consequences of poor interfacial bonding have been widely documented and will be discussed in subsequent chapters. Figure 2.9 [17] illustrates evidence of poor bonding where the SEM images show clean surfaces of the pulled-out fibers. These images are, however, not typical of resin-infused composites. They are for polyvinylidene chloride reinforced with glass fibers at different fiber volume fractions and produced by compression molding.

2.4   PREPREG-BASED AUTOCLAVE CURING

Fibers of carbon, glass, or aramid pre-impregnated with a resin and formed as a tape, called a prepreg, has been a key technology initially in aerospace parts, and increasingly in other applications. Here the prepreg-based manufacturing process with a thermosetting resin using an autoclave will be reviewed. The review is not meant to cover all aspects of this process, but only to provide enough background to understand the nature of manufacturing defects that play a role in failures. For a comprehensive review of the autoclave curing process, see [3].

The prepreg-based autoclave curing starts by stacking prepreg tapes onto one another in desired fiber orientations. The prepregs contain thin layers of straight

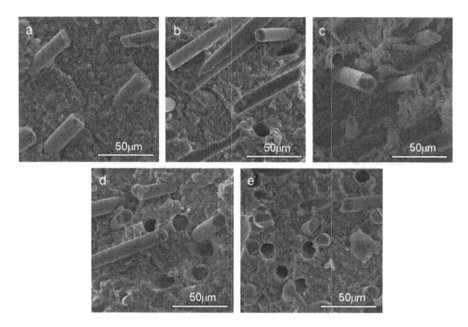

FIGURE 2.9 SEM images of a glass fiber–reinforced polyvinylidene chloride made by compression molding. Images a–e are for increasing fiber volume fraction. Source: Zhang, Y.H., Bai, B.F., Li, J.Q., Chen, J.B. and Shen, C.Y., 2011. Multifractal analysis of the tensile fracture morphology of polyvinylidene chloride/glass fiber composite. *Applied Surface Science*, *257*(7), 2984–2989.

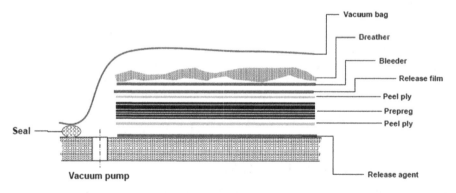

FIGURE 2.10 Details of the vacuum bagging for autoclave curing. Source: Fernlund, G., Mobichon, C., Zobeiry, N., 2018. Autoclave Processing, Comprehensive Composite Materials II, Vol. 2, 42–62.

fibers or fabrics in a semi-cured resin (stage-B cured) with a protective layer called peel ply on each side that is peeled off before stacking the prepreg tape onto another tape. The tack of the semi-cured resin allows stable placement of the tapes as layers of laminate. After completing the prepreg layup of the part, it is vacuum bagged, as illustrated in Figure 2.10. A release agent between the part and the tool allows the separation of the part from the tool after completion of curing. Similarly, a release film between the part and the bleeder allows separating the two after curing. A breather is used to protect the vacuum bag and to allow air to pass through.

The vacuum-bagged part is then placed in an autoclave. Figure 2.11 shows a sketch of a common autoclave. The autoclave is essentially a pressure vessel with controlled heating. The hot air circulation assures a uniform temperature, which is varied as specified by the prepreg manufacturer to achieve proper curing of the part. Typical cure temperatures for epoxies are in the range of 120–180°C, and the autoclave pressures lie between 3 and 7 bars. After curing, the part is separated from the bleeder and the tool with the help of the release films. Finally, the peel plies are removed resulting in a clean part.

2.4.1 Manufacturing Defects Induced by Autoclave Curing

Although the prepreg tape is generally of high quality in terms of fiber orientation accuracy and low void content, there are other sources of defects in the autoclave curing of composite parts. The following gives an overview of the formation of these defects.

2.4.1.1 Spring-Back

Spring-back is used to describe the permanent deviation of a part from its designed dimensions on removal from the mold. This is caused by internal stresses that develop during the curing process. These stresses are due to three main mechanisms [18]: differential thermal expansion, cure shrinkage, and part–tool interaction. The thermal expansion mismatch at the constituent level generates residual stresses in fibers

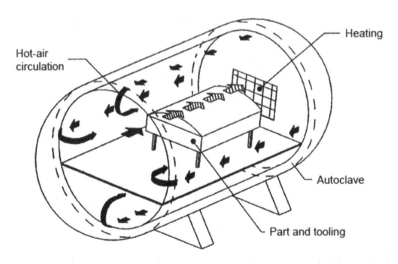

FIGURE 2.11   Details of the vacuum bagging for autoclave curing. Source: Fernlund, G., Mobichon, C., Zobeiry, N., 2018. Autoclave Processing, Comprehensive Composite Materials II, Vol. 2, 42–62.

and matrix during cooldown from the cure temperature. In a laminate with differently oriented plies, the thermal expansion anisotropy of the plies induces residual stresses in the plies. This results in various distortions of the laminate depending on the laminate geometry and layup. Cure shrinkage is a characteristic of polymers caused by chemical reactions in the curing process. Finally, the temporary bonding between the surfaces of a part and the tool on which it is in contact during the autoclave curing can produce significant stresses in the part in laminate layers close to the tool. All these effects are mainly undesirable distortions of the part geometry, but they can also cause failure if the residual stresses add onto the imposed stresses under service loading.

2.4.1.2   Void Formation

A well-controlled prepreg manufacturing process reduces the formation of voids due to air entrapment. However, during autoclave curing the release of volatiles and moisture from the resin can be a source of void formation within the prepreg layers. The size and shape of these intralaminar voids depend on the temperature and pressure during curing. The moisture-induced void formation has been studied by [19] where a significant effect of pressure was found. The effect of pressure in the autoclave on void migration and collapse is illustrated in Figure 2.12 [3]. The heat applied during the curing process reduces the viscosity of the resin, trapping the bubbles formed by the release of the volatiles and moisture. The pressure then collapses some bubbles and drives others to other locations in the part.

Another source of void formation in the autoclave curing process is the entrapped air between prepregs during the layup process. Applying a vacuum can draw such voids out of the laminate, but if the voids are squeezed under pressure, they can stay

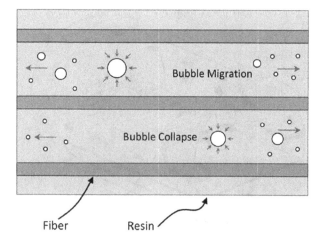

FIGURE 2.12   The effect of pressure in the autoclave on collapse and migration of the bubbles (voids) formed by release of volatiles and moisture in a prepreg. Source: Fernlund, G., Mobichon, C., Zobeiry, N., 2018. Autoclave Processing, Comprehensive Composite Materials II, Vol. 2, 42–62.

within the laminate and form dry spots. An example of voids trapped between two plies in a laminate is shown in Figure 2.13 [20] where a vacuum was not used in the autoclave curing of a laminate. As evidenced, the consolidation pressure in the autoclave flattened the voids that could not migrate because of lack of vacuum.

2.4.1.3 Defects in Complex-Shaped Parts

When the geometry of a part is complex, e.g., consisting of sharp changes of curvature, corners, and inserts, the resin infusion process causes mold-filling difficulties resulting in defects. Therefore, a prepreg-based method is preferred in such cases. However, other issues arise that cause defects. Complex shape molding with autoclave has been examined from a defects point of view, see a recent review [21]. Figure 2.14 is taken from this review and illustrates the type of manufacturing defects that can result at the corner of a part. Other situations giving rise to such defects are stiffeners and tapering of a part. In thick composite parts, defects result from non-uniform temperature and pressure variations through the thickness.

2.4.1.4 Defects in Automated Tape Laying (ATL)

Prepreg laying on a tool is mostly done manually by trained personnel in a clean room. The nature and degree of defects in this case are difficult to control and assess, particularly for large parts. Increasingly, therefore, and for cost reduction reasons, machines have been developed for automated tape laying (ATL). A historical review of machines from the 1960s to the early years of the 21st century is given in [22]. Figure 2.15 shows the essential details of the ATL process [22]. Certain defects that

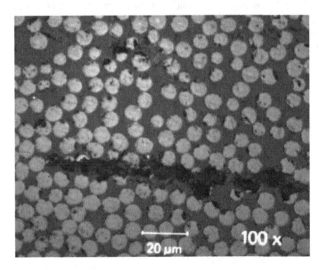

FIGURE 2.13 Voids trapped between two plies in an autoclave-cured laminate where the vacuum was not applied. Source: Huang, Y., Varna, J. and Talreja, R., 2011. The effect of manufacturing quality on transverse cracking in cross ply laminates. In *International Conference on Processing and Fabrication of Advanced Materials: 15/01/2011-18/01/2011* (pp. 552-559). Centre for Advanced Composite Materials, University of Auckland.

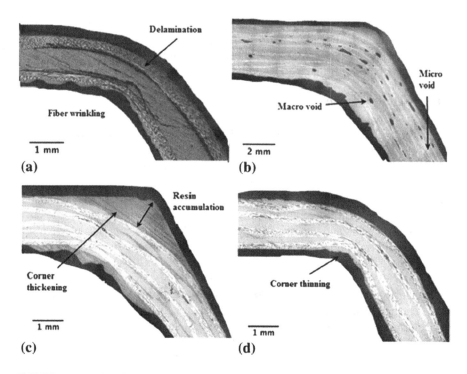

FIGURE 2.14 Manufacturing defects resulting from prepreg-based autoclave curing of a part near a corner. (a) Delamination, (b) voids, (c) resin accumulation, and (d) resin squeezing. Source: Hassan, M.H., Othman, A.R. and Kamaruddin, S., 2017. A review on the manufacturing defects of complex-shaped laminate in aircraft composite structures. *The International Journal of Advanced Manufacturing Technology*, 91, 4081–4094.

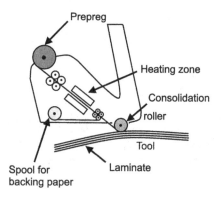

FIGURE 2.15 A schematic description of the automated tape laying (ATL) process. Source: Dirk, H.J.L., Ward, C. and Potter, K.D., 2012. The engineering aspects of automated prepreg layup: History, present and future. *Composites Part B: Engineering*, 43(3), 997–1009.

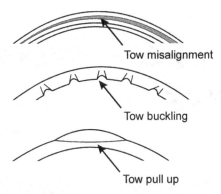

FIGURE 2.16　Common defects in automated placement of fibers on a tool. Source: Dirk, H.J.L., Ward, C. and Potter, K.D., 2012. The engineering aspects of automated prepreg layup: History, present and future. *Composites Part B: Engineering, 43*(3), 997–1009.

are inherent in the automated prepreg laying on a tool are misalignment, buckling, and the pull-up of fibers, as sketched in Figure 2.16 [22].

2.5　OUT-OF-AUTOCLAVE (OOA) PREPREG PROCESSING

In the conventional prepreg-based autoclave curing described above, a substantial cost lies in the use of an autoclave. Efforts have therefore been made to develop curing processes that would not require the use of a pressurized vessel. Such processes have come to be known as out-of-autoclave (OOA) curing processes. In addition to reducing the cost of manufacturing by not using an autoclave, these processes allow the manufacturing of larger parts that are not limited by the available size of an autoclave. A comprehensive review of the OOA curing processes is given by [4]. Here the focus is on the manufacturing defects that can result in the OOA curing processes with a view to their roles in initiating failure of composite parts.

In place of a pressurized vessel used for heating in the autoclave curing, a convection oven is often used in OOA curing. The temperature distribution in that case is not very uniform due to the laminate and the tool acting as heat sinks. The curing of the resin can be affected by this, leading to regions of partially cured resin. An alternative way is to use heated tools and/or heat blankets. The heat conduction in this case allows shorter curing time but control of heat losses is necessary to avoid spatial gradients of temperature.

Void formation in OOA-processed parts is of major concern. The void collapse and void migration mechanisms illustrated in Figure 2.12 for the case of autoclave processing are not available to the same extent in OOA processing as the only pressure for consolidation comes from the applied vacuum. Figure 2.17 illustrates the steps in achieving a void-free part in an OOA process. Note that removal of the voids that are formed by air entrapment and other means are evacuated by vacuum only as additional pressure is not available for collapsing and migrating the voids.

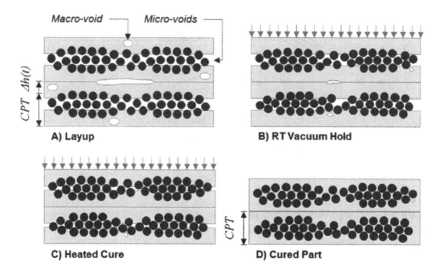

FIGURE 2.17 Steps in the OOA curing process for achieving a void-free microstructure of the cured part. CPT refers to cured ply thickness, which reduces in time as the resin shrinks during cure and the entrapped air is evacuated. Source: Hubert, P., Centea, T., Grunefelder, L., Nutt, S., Katz, J., Levy, A., 2018. Out-of-autoclave Prepreg Processing. *Comprehensive Composite Materials II*, 2, 63–94.

One development related to OOA curing is prepregs that allow the manufacturing of parts without high consolidation pressure normally used in autoclaves. These prepregs are known as vacuum bag only (VBO) prepregs. In the development of these prepregs, measures are taken to allow the removal of entrapped air and volatiles at low pressures. One pathway is to use perforated resin films in prepregs that create gaps through which the gasses can be drawn out by vacuum. Another method, more commonly used, is to do partial impregnation of the prepregs leaving dry zones in the prepregs formed evacuation channels that facilitate the flow of gasses during vacuum consolidation. In both these methods, the gaps and dry zones are expected to be filled by the resin as curing takes place. The cure kinetics and viscosity evolution during curing are important factors in the successful avoidance of void formation. An additional challenge in the development of the VBO prepregs is to account for the so-called out-time, i.e., the time the prepreg is at room temperature before curing begins. The polymerization/crosslinking of the resin during out-time can affect the curing process during the manufacture of parts. This effect can be significant in large parts that can take weeks to complete.

A recent XCT study reported details of the dry channels at the beginning and at different stages during the VBO curing process [23]. Figure 2.18 from that study shows tomographs of a laminate and of a region of a dry channel. The degree to which such dry channels are effective in vacuum evacuation of the entrapped air and volatiles from the resin, and the extent to which the channels are filled with resin, determine the voids in the cured part.

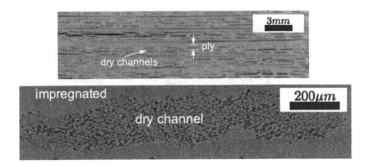

FIGURE 2.18 A XCT tomographs of a laminate (top) and of a region of a dry channel (bottom). Source: Torres, J.J., Simmons, M., Sket, F. and González, C., 2019. An analysis of void formation mechanisms in out-of-autoclave prepregs by means of X-ray computed tomography. *Composites Part A: Applied Science and Manufacturing, 117,* 230–242.

FIGURE 2.19 Main steps in a thermoplastic thermoforming process. Source: Xiong, H., Hamila, N. and Boisse, P., 2019. Consolidation modeling during thermoforming of thermoplastic composite prepregs. *Materials, 12*(18), 2853.

2.6 THERMOFORMING

Thermoforming is suitable mainly for thermoplastics-based composites. It is a process that allows the rapid transformation of flat laminates into complex shapes by the application of heat and pressure. The steps involved in the process are illustrated in Figure 2.19 [24]. The quality of a manufactured part by thermoforming depends on the many process parameters that have different degrees of influence on the quality. Efforts have been made to optimize the process, most recently by machine learning methods [25]. Among the process parameters are bulk-forming temperature, mold temperature, cooling rate, dwell time, and forming pressure. Additionally, the laminate fiber architecture, ply layup, and the geometry of the part influence the part quality. Thermoplastics can also change their flow properties significantly due to heating and cooling rates. An example of this is polyetheretherketone (PEEK) which changes its crystallinity and therefore its deformation response.

An illustration of how wrinkles result in a thermoforming process was reported in [26] where seven layers of 5-harness satin carbon/PEEK prepreg layers were thermoformed into a bowl shape. Figure 2.20(a) displays a schematic of the process showing the die, the stacked preform, and the punch. Figure 2.20(b) shows an image of the final product and Figure 2.20(c) indicates the wrinkles formed in a corner region of the bowl. A numerical simulation of the thermoforming process suggested that the wrinkles were primarily the result of non-uniform temperatures during the process.

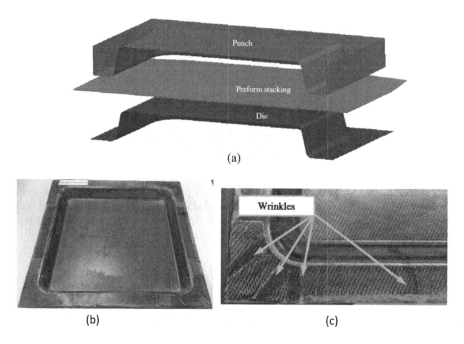

FIGURE 2.20 (a) Schematic description of the thermoforming process showing the die, preform, and punch, (b) the finished part, and (c) details of the wrinkles at the end of the forming process. Source: Wang, P., Hamila, N. and Boisse, P., 2013. Thermoforming simulation of multilayer composites with continuous fibres and thermoplastic matrix. *Composites Part B: Engineering, 52,* 127–136.

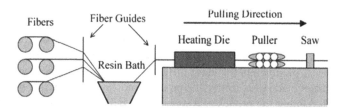

FIGURE 2.21 Schematic of the pultrusion process. Source: Baran, I., Tutum, C.C., Nielsen, M.W. and Hattel, J.H., 2013. Process-induced residual stresses and distortions in pultrusion. *Composites Part B: Engineering, 51,* 148–161.

2.7 PULTRUSION

This process is widely used for the cost-effective manufacturing of parts with a constant cross section over a length, and to a limited extent when the cross section varies gradually. Examples of parts made by the process are beams for buildings and bridges and bumpers in the automotive industry. A simple schematic of the process is shown in Figure 2.21. Fibers from a creel are guided through a resin impregnation unit and then pulled through a forming die where consolidation and curing take place. Many variations of this basic process have been developed, e.g., by using fabrics and preforms to produce a variety of profiles of solid and hollow cross sections, and by injecting the resin in a preform before pulling it through the die. The process parameters that govern the quality of the pultruded composite consist of pulling speed, curing temperature, resin reaction kinetics, etc., as well as fiber architecture. The temperature variation during the process is especially important if a thermoplastic resin is used. One of the important consequences of the process is the development of residual stresses causing geometrical distortions of the part. A modeling study of this effect is reported in [27].

2.8 FILAMENT WINDING

In its classical version, the filament winding process consists of a rotating mandrel and a translationally moving winding unit as sketched in Figure 2.22 [28]. The fibers can be impregnated with liquid resin just before winding, in which case it is called wet winding, or it can be impregnated after winding, known as dry winding. The parts produced have rotationally symmetric geometry, the common examples of which are drive shafts, pipes, and storage tanks. The modern versions of the filament winding are based on robot-guided winding and detachable mandrels to make complex shapes of parts possible [28].

After the completion of the winding process, the mandrel with the fibers and resin moves into an oven or an autoclave for curing. The mandrel is then removed or in some cases becomes part of the composite structure. In some cases of complex-shaped parts, the mandrel is made of a soluble material that is dissolved and washed out of the part. Mandrels can also be collapsible or made of smaller pieces that are removed after curing.

Filament winding is also used for thermoplastic (narrow) tapes, in which case the resin bath is not needed. The material is heated while being wound onto the mandrel, compacted, consolidated, and cooled in a single operation.

The nature of defects resulting from filament winding depends on the pre-tensioning of fibers or tapes and the degree of control of the winding process. Common

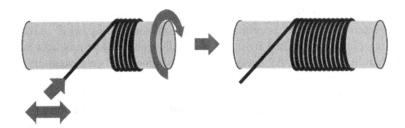

FIGURE 2.22 Schematic of the classical filament winding using a rotating mandrel and a translationally moving unit. Source: Fleischer, J., Teti, R., Lanza, G., Mativenga, P., Mohring, H.-C., Caggiono, A., 2018, Composite materials parts manufacturing. *CIRP Annals – Manufacturing Technology*, *67*, 603–626.

defects are dry interfaces, locally misaligned fibers, varying fiber volume fractions, and matrix voids.

2.9   ADDITIVE MANUFACTURING (AM)

Additive manufacturing (AM), also known as 3D printing, is a rapidly growing field. It uses computer-aided design (CAD) software or 3D object scanners to direct layer-wise deposition of material to create precise geometric shapes. In contrast to the traditional shaping of metallic and ceramic structural parts, where operations such as milling, machining, carving, etc. are conducted, AM creates an object by adding material. The layering can be done with a variety of materials, e.g., metals, ceramics, thermoplastics, and composites. Several AM techniques have been developed to conduct the layering process. Brief descriptions of the common techniques follow.

In the **stereolithography** (SLA) process, the part model is built on a platform positioned just below the surface in a vat of liquid resin. A low-powered ultraviolet (UV) laser, programmed with the previously created CAD slice data, traces out the first layer of the part with its highly focused UV light beam, scanning and curing the resin within the boundaries of the slice outline until the entire area within the slice cross section is solidified. An elevator then incrementally lowers the platform into the liquid polymer to a depth equal to the slice thickness, and a sweeper recoats the solidified layer with the liquid polymer. Following this, the laser traces out a second layer on top of the first. The process is repeated until the part is complete. After removal from the vat, supports are removed from the part, which then is placed in a UV oven for additional curing.

In **fused deposition modeling** (FDM), a 3D object is built one layer at a time. A polymer filament is unwound from a coil, supplying material to a heated extrusion nozzle, which controls the flow. The nozzle is mounted over a mechanical stage and can be moved horizontally and/or vertically. The nozzle deposits a thin bead of extruded polymer on a supporting material. Successive extruded layers bond with the previous layers, and then solidify immediately. The entire system is contained in a chamber held at a temperature just below the polymer melting point. No postprocessing is required after the part is removed from the chamber.

The bonding between layers in the FDM process occurs differently in amorphous and semi-crystalline polymers. In amorphous polymers, thermal fusion, and polymer interdiffusion are the primary mechanisms for bonding, while in the semi-crystalline polymers bonding by co-crystallization occurs and produces stronger bonds than in amorphous polymers. Filaments of amorphous and semi-crystalline thermoplastics are commercially available for FDM but may not produce sufficient stiffness and strength properties needed for some parts. Reinforcing these filaments with short carbon fibers has been attempted [29–31]. Carbon fibers powdered to a length of 0.1–0.15 mm have been combined with pellets of acrylonitrile butadiene styrene (ABS) at 3–15 weight% of carbon to produce filaments as feedstock for FDM. Results show that porosity levels can become unacceptable (>2%) for carbon loading of 10% or higher. This limits the property enhancement that can be achieved by the addition of short carbon fibers. However, attempts have been made to add longer carbon fibers, 0.2–0.4 mm in length, and orient them (~90%) to achieve higher properties in the fiber direction [32]. Unfortunately, the porosity problems remain.

In the **selective laser sintering** (SLS) process, heat from a CO_2 laser is used to process a variety of materials in powder rather than liquid form, including nylon, and glass fiber- or carbon fiber-filled nylons. In an enclosed unit, a CO_2 laser and a mirrored reflector system are mounted over a build table or pedestal, which supports the part. A roller distributes a thin layer of powdered material over the pedestal surface, and then the mirror system directs the laser beam onto the powder layer. As the beam scans back and forth across the material, the laser turns on and off, selectively sintering the powder in a pattern identical in size and shape to the cross-sectional slice derived from the converted CAD file. The pedestal is then lowered the distance of the layer thickness, another layer of powder is rolled over the cooled and now solidified first layer, and the sintering process is repeated, bonding the second layer to the first. The process repeats, in layers of 0.08–0.15 mm thickness, until the part is complete.

Digital light processing (DLP), like the stereolithography platforms, uses light-curable resins but can process them faster using a continuous process rather than incremental layering that involves projecting the entire image onto a liquid photopolymer bath rather than scanning over successively applied layers of powdered or liquid resin with a point energy source or depositing layers of material and applying heat.

The process with the most potential for making AM parts appears to be the FDM process. A recent review [33] renamed it the fused filament fabrication (FFF) approach and focused on its advances and future prospects with respect to producing composite parts. It is recognized that the process implementation is largely empirical with inadequate tools for simulation. Most of the advances are in short-fiber reinforced composites, and progress is slowly being made to address placing longer fibers in the filaments. One issue is determining the fiber orientation in the material as it flows through the die and is deposited onto the printing bed. The flow-direction viscosity affects the fiber orientation, which in turn influences the viscosity.

2.9.1 Defect Formation

In all AM processes used for composite manufacturing, the major issues related to the part quality are the accuracy of fiber orientation, the uniformity of fiber volume fraction, adequate adhesion between layers, and the presence of matrix voids. Since the simulation tools for the AM processes are lacking, predictions of the defects are not yet possible. Assessments of the AM processes appear to rely on overall stiffness and strength properties, while initiation and progression of local failure, and its criticality remain to be understood.

2.10 POST-PROCESSING IN PARTS MANUFACTURING

The manufacturing processes described above do not always produce parts ready to be placed in service. Often, therefore, additional steps are taken beyond those processes. These consist of machining and joining for the most part. The machining can involve cutting, milling, drilling, and trimming by conventional machines and with advanced techniques such as laser beam machining and water jet machining. The joining methods are many, and their selection depends on the cost and size/shape of the parts. Thus, joining can be done by mechanical fasteners or by adhesives. For thermoplastics-based composites, thermal joining is possible where the parts are pressed together and heated to the melting temperature of the thermoplastic and cooled. Heating can be achieved by laser beams, the advantage being local heating at high rates. For a review of post-processing techniques in composite manufacturing [28] can be consulted.

Each step in the composite manufacturing process involves introducing defects in the composite that can potentially become a source of failure initiation and progression under service conditions. The post-processing steps of machining and joining introduce local defects that are not distributed within the volume of a composite but can still produce damage that can spread into the composite volume. In any case, the structural integrity and durability are affected.

2.11 CONCLUSION

This chapter has described various common manufacturing methods used for structures made of PMCs, such as liquid composite molding, prepreg-based autoclave curing, out-of-autoclave prepreg processing, automated tape laying, thermoforming, pultrusion, filament winding, and the more recent, additive manufacturing. For each method, the sources of defects have been discussed as they relate to fibers, matrix, and interfaces. Geometrical defects induced by autoclave curing such as spring-back and automated tape laying have also been treated. Void formation in the polymer matrix has been discussed as it pertains to different resin infusion and curing processes. Finally, post-processing such as machining and joining has also been described. Knowledge of manufacturing processes is essential in developing design approaches that can make use of composite materials effectively by attending to the defects that are inherent in practical manufacturing. The focus of this book is the incorporation of manufacturing defects in analyses to prevent failure in service environments. For that purpose, this chapter has provided relevant background.

REFERENCES

1. Hoa, S.V., 2018. Manufacturing of composites – An overview. *Comprehensive Composite Materials II*, 2, pp. 1–23.
2. Advani, S.G., 2018. Role of process models in composites manufacturing. *Comprehensive Composite Materials II*, 2, pp. 24–41.
3. Fernlund, G., Mobichon, C. and Zobeiry, N., 2018. Autoclave processing. *Comprehensive Composite Materials II*, 2, pp. 42–62.
4. Hubert, P., Centea, T., Grunefelder, L., Nutt, S., Katz, J. and Levy, A., 2018. Out-of-autoclave prepreg processing. *Comprehensive Composite Materials II*, 2, pp. 63–94.
5. Bodaghi, M., Lomov, S.V., Simacek, P., Correia, N.C. and Advani, S.G., 2019. On the variability of permeability induced by reinforcement distortions and dual scale flow

in liquid composite moulding: A review. *Composites Part A: Applied Science and Manufacturing, 120*, pp. 188–210.

6. Hindersmann, A., 2019. Confusion about infusion: An overview of infusion processes. *Composites Part A: Applied Science and Manufacturing, 126*, p. 105583.

7. Shojaei, A., Ghaffarian, S.R. and Karimian, S.M.H., 2003. Modeling and simulation approaches in the resin transfer molding process: A review. *Polymer Composites, 24*(4), pp. 525–544.

8. Goren, A. and Atas, C., 2008. Manufacturing of polymer matrix composites using vacuum assisted resin infusion molding. *Archives of Materials Science and Engineering, 34*(2), pp. 117–120.

9. Van Oosterom, S., Allen, T., Battley, M. and Bickerton, S., 2019. An objective comparison of common vacuum assisted resin infusion processes. *Composites Part A: Applied Science and Manufacturing, 125*, p. 105528.

10. Song, Y.S., 2006. Characterization of mold filling in vacuum infusion process. *Advanced Composites Letters, 15*(2), p. 096369350601500203.

11. Lopatnikov, S., Simacek, P., GillespieJr, J. and Advani, S.G., 2004. A closed form solution to describe infusion of resin under vacuum in deformable fibrous porous media. *Modelling and Simulation in Materials science and Engineering, 12*(3), p. S191.

12. Wang, P., Drapier, S., Molimard, J., Vautrin, A. and Minni, J.C., 2012. Numerical and experimental analyses of resin infusion manufacturing processes of composite materials. *Journal of Composite Materials, 46*(13), pp. 1617–1631.

13. Pierce, R.S. and Falzon, B.G., 2017. Simulating resin infusion through textile reinforcement materials for the manufacture of complex composite structures. *Engineering, 3*(5), pp. 596–607.

14. Hamidi, Y.K. and Altan, M.C., 2017. Process-induced defects in liquid molding processes of composites. *International Polymer Processing, 32*(5), pp. 527–544.

15. Mehdikhani, M., Gorbatikh, L., Verpoest, I. and Lomov, S.V., 2019. Voids in fiber-reinforced polymer composites: A review on their formation, characteristics, and effects on mechanical performance. *Journal of Composite Materials, 53*(12), pp. 1579–1669.

16. Lightfoot, J.S., Wisnom, M.R. and Potter, K., 2013. Defects in woven preforms: Formation mechanisms and the effects of laminate design and layup protocol. *Composites Part A: Applied Science and Manufacturing, 51*, pp. 99–107.

17. Zhang, Y.H., Bai, B.F., Li, J.Q., Chen, J.B. and Shen, C.Y., 2011. Multifractal analysis of the tensile fracture morphology of polyvinylidene chloride/glass fiber composite. *Applied Surface Science, 257*(7), pp. 2984–2989.

18. Wisnom, M.R., Gigliotti, M., Ersoy, N., Campbell, M. and Potter, K.D., 2006. Mechanisms generating residual stresses and distortion during manufacture of polymer–matrix composite structures. *Composites Part A: Applied Science and Manufacturing, 37*(4), pp. 522–529.

19. Grunenfelder, L.K. and Nutt, S.R., 2010. Void formation in composite prepregs–Effect of dissolved moisture. *Composites Science and Technology, 70*(16), pp. 2304–2309.

20. Huang, Y., Varna, J. and Talreja, R., 2011. The effect of manufacturing quality on transverse cracking in cross ply laminates. In *International Conference on Processing and Fabrication of Advanced Materials: 15/01/2011-18/01/2011* (pp. 552–559). Centre for Advanced Composite Materials, University of Auckland.

21. Hassan, M.H., Othman, A.R. and Kamaruddin, S., 2017. A review on the manufacturing defects of complex-shaped laminate in aircraft composite structures. *The International Journal of Advanced Manufacturing Technology, 91*, pp. 4081–4094.

22. Dirk, H.J.L., Ward, C. and Potter, K.D., 2012. The engineering aspects of automated prepreg layup: History, present and future. *Composites Part B: Engineering, 43*(3), pp. 997–1009.

23. Torres, J.J., Simmons, M., Sket, F. and González, C., 2019. An analysis of void formation mechanisms in out-of-autoclave prepregs by means of X-ray computed tomography. *Composites Part A: Applied Science and Manufacturing, 117*, pp. 230–242.
24. Xiong, H., Hamila, N. and Boisse, P., 2019. Consolidation modeling during thermoforming of thermoplastic composite prepregs. *Materials, 12*(18), p. 2853.
25. Nardi, D. and Sinke, J., 2021. Design analysis for thermoforming of thermoplastic composites: Prediction and machine learning-based optimization. *Composites Part C: Open Access, 5*, p. 100126.
26. Wang, P., Hamila, N. and Boisse, P., 2013. Thermoforming simulation of multilayer composites with continuous fibres and thermoplastic matrix. *Composites Part B: Engineering, 52*, pp. 127–136.
27. Baran, I., Tutum, C.C., Nielsen, M.W. and Hattel, J.H., 2013. Process-induced residual stresses and distortions in pultrusion. *Composites Part B: Engineering, 51*, pp. 148–161.
28. Fleischer, J., Teti, R., Lanza, G., Mativenga, P., Mohring, H.-C. and Caggiono, A., 2018. Composite materials parts manufacturing. *CIRP Annals – Manufacturing Technology, 67*, pp. 603–626.
29. Ning, F., Cong, W., Qiu, J., Wei, J. and Wang, S., 2015. Additive manufacturing of carbon fiber reinforced thermoplastic composites using fused deposition modeling. *Composites Part B: Engineering, 80*, pp. 369–378.
30. Zhang, W., Cotton, C., Sun, J., Heider, D., Gu, B., Sun, B. and Chou, T.W., 2018. Interfacial bonding strength of short carbon fiber/acrylonitrile-butadiene-styrene composites fabricated by fused deposition modeling. *Composites Part B: Engineering, 137*, pp. 51–59.
31. Tekinalp, H.L., Kunc, V., Velez-Garcia, G.M., Duty, C.E., Love, L.J., Naskar, A.K., Blue, C.A. and Ozcan, S., 2014. Highly oriented carbon fiber–polymer composites via additive manufacturing. *Composites Science and Technology, 105*, pp. 144–150.
32. Vidya, K., Chen, X., Ajinjeru, C., Hassen, A.A., Lindahl, J.M., Failla, J., Kunc, V. and Duty, C.E., 2016. Additive manufacturing of high performance semicrystalline thermoplastics and their composites. OSTI.gov.
33. Brenken, B., Barocio, E., Favaloro, A., Kunc, V. and Pipes, R.B., 2018. Fused filament fabrication of fiber-reinforced polymers: A review. *Additive Manufacturing, 21*, pp. 1–16.

3 Quantification of Manufacturing Defects

3.1 INTRODUCTION

As described in Chapter 1, the defects in composite structures resulting from a manufacturing process depend on the process parameters as well as on the post-processing operations conducted before deploying the structure in service. Chapter 2 was devoted to describing the common manufacturing processes for PMCs and the nature of defects induced by those processes. Yet, it is not possible to predict the defects by modeling and simulation, while efforts in this direction continue. The necessary approach currently is therefore to observe the defects by non-destructive techniques, if possible, or by sectioning the composite and directly observing the defects by optical microscopy, scanning electron microscopy (SEM), etc. In any case, the defects are found to be non-uniform both in their geometry and in their spatial distribution. For assessing their effects on materials performance, therefore, stochastic descriptors of the defects must be constructed. This chapter will discuss approaches to developing such descriptors for defects that will help define the real initial material state (RIMS) described in Chapter 1.

3.2 NON-UNIFORMITY OF FIBER DISTRIBUTION IN THE COMPOSITE CROSS SECTION

When cross sections normal to fibers in composite materials are examined, one finds non-uniformity in fiber distribution as well as in the fiber diameter. An example is shown in Figure 3.1, where an SEM micrograph along with its processed image is shown [1]. The areas where fibers are clustered and where resin pockets exist are visible. Variations in fiber diameter can also be seen.

These deviations from the desired (ideal) uniform fiber patterns have consequences on the failure behavior that will be discussed in a later chapter. Here, we shall focus on how to generate stochastic descriptors that can be taken to represent the microstructural morphology.

A general approach to describing the microstructural morphology of heterogeneous solids with a view to determining their macroscopic properties has been treated in [2]. Our interest in this book is to focus on how the microscopic morphological features affect failure in fiber-reinforced composites. From this perspective, the work presented by Pyrz [3–4] is of interest. In [3] Pyrz presented a statistical analysis of fiber distribution in a composite cross section such as that illustrated in Figure 3.1 by employing a second-order intensity function proposed by Ripley [5]

DOI: 10.1201/9781003225737-3

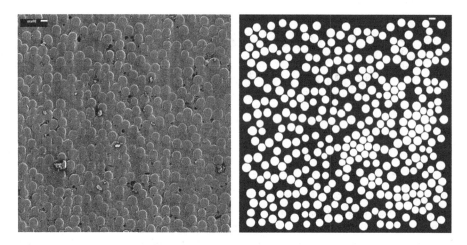

FIGURE 3.1   A SEM micrograph (a) and a processed image (b) of a cross section of a glass/
epoxy composite. Source: Grufman, C. and Ellyin, F., 2007. Determining a representative volume element capturing the morphology of fibre reinforced polymer composites. *Composites
Science and Technology*, *67*(3-4), 766_775.

and using Dirichlet tessellation. The variation in the size and shape of fibers was
not considered, and the fibers were replaced by their centroids. Based on a reasonable assumption that local stress fields in the matrix are primarily influenced by
inter-fiber spacing, the statistical analysis to develop descriptors of microstructural
morphology was aimed at that spacing. The procedure developed in [3] allows the
classification of the fiber patterns as well as for quantifying the inter-fiber distances.
It is described as follows.

Consider a field of observation such as the image shown in Figure 3.1 (b). Let
there be N distinct fibers (replaced by their centers). To each fiber, a "zone of influence" is assigned following the procedure for constructing the so-called Voronoi
polygons (also called Dirichlet tessellations) sketched in Figure 3.2 [1]. As illustrated
in the figure, such polygons are constructed by connecting two neighboring fiber
centers by straight lines and then drawing normal to these lines at points midway
between the fiber centers. These lines intersected by other such lines corresponding
to a given fiber form a Voronoi polygon. The area within such a polygon (the hatched
area in Figure 3.2) is viewed as the zone of influence of the considered fiber. The
numerical algorithms for constructing Dirichlet tessellations for a large number of
fibers are described in [6].

Figure 3.3 illustrates the calculation of Ripley's K-function [5], which is given by,

$$K(r) = \frac{A}{N^2} \sum_{k=1}^{N} \frac{I_k(r)}{w_k} \qquad (3.1)$$

where N is the number of points (fiber centers) in the observation area A, $I_k(r)$ is the
number of points within the circle of radius r centered at a considered (arbitrary)

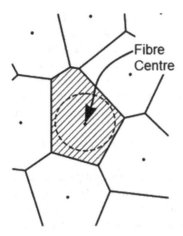

FIGURE 3.2 Illustration of Voronoi polygons showing the polygon (shaded area) for the circular fiber indicated Source: Grufman, C. and Ellyin, F., 2007. Determining a representative volume element capturing the morphology of fibre reinforced polymer composites. *Composites Science and Technology, 67*(3-4), 766–775.

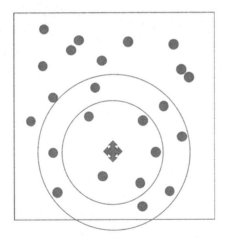

FIGURE 3.3 A square window of observation containing a nonuniform distribution of fibers. Two circles centered at an arbitrary fiber are shown, the outer circle going beyond the field of observation.

point, excluding that point, and w_k is a correction factor for a circle that goes beyond an edge (such as the outer circle in Figure 3.3) and is equal to the ratio of the circumference of the circle within A to the total circumference of that circle.

Pyrz [4] described two extreme cases of point distribution patterns that are not realized in practical manufacturing. One is the completely random pattern defined by the region where each possible location is equally likely to be occupied by a fiber and all locations are independent of one another. The probability distribution of inter-point distance in this case is given by the Poisson distribution. Although theoretically possible for points, this case is not realizable for fibers of non-zero diameters. The other extreme of uniform point distribution pattern is also not possible in real manufacturing. The cases between the extremes were classified by Pyrz [4] as clusters and hard-core patterns. Figure 3.4 illustrates the triple and single cluster patterns as well as a hard-core pattern and a uniform pattern. The hard-core pattern is constructed by specifying a minimum permissible distance between points in an otherwise random distribution.

Plotting the $K(r)$-function (Equation 3.1) against the variable r gives a characteristic curve for each fiber distribution pattern. The K-function for the baseline Poisson distribution pattern is πr^2 and the clustered patterns deviate increasingly from this curve as the number of clusters increases. The deviation of the hard-core distribution pattern depends on the minimum permissible inter-point distance. The K-function of the regular pattern shows discontinuities at inter-point distances indicating empty spaces at those distances.

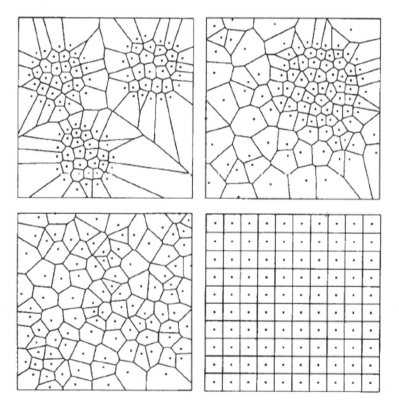

FIGURE 3.4 Point distribution patterns: triple and single clusters (top), and hard-core and regular patterns (bottom). Source: Pyrz, R., 1994. Correlation of microstructure variability and local stress field in two-phase materials. *Materials Science and Engineering: A*, *177*(1–2), 253–259.

While the K-function provides a useful way to discriminate between patterns in the fiber center distributions, the inter-fiber spacing correlates with another function, called the pair distribution function $g(r)$ [4]. This function is obtained by observing that the average number of fiber centroids within a ring of thickness dr that lies between two concentric circles of radii r and $r + dr$, centered at a given fiber centroid, is given by $dK(r)$. Dividing this value by the area of the ring, and further dividing it by the number of fiber centroids per unit area of the observation window, gives,

$$g(r) = \frac{A}{2\pi N} \frac{dK(r)}{dr} \qquad (3.2)$$

This function provides the intensity of inter-fiber distances and is a relevant statistic for local stress fields. When plotted against the radial distance r, $g(r)$ attains local maxima at the most frequent occurrences and minima at the least frequent occurrences of the inter-fiber distances, respectively. For the Poisson random process, in which the occurrences of events are independent of one another, $g(r) = 1$, while for more frequent occurrences than expected in the Poisson process, $g(r) > 1$, and for less frequent occurrences, $g(r) < 1$.

One example of the $K(r)$-function and the $g(r)$-function calculated from SEM micrographs is shown in Figure 3.5 [7]. The corresponding images of observed fiber distribution are shown in Figure 3.6.

3.2.1 REPRESENTATION OF NON-UNIFORM FIBER DISTRIBUTION IN A CROSS SECTION

For modeling the effects of defects, it is important to have descriptors that contain sufficient representation of the non-uniformity of fiber distribution in a composite cross section. What constitutes a sufficient representation depends on the purpose of modeling, e.g., estimation of average properties such as elastic moduli or initiation

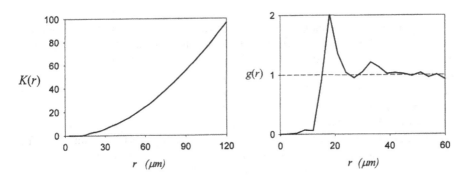

FIGURE 3.5 The K- and g-functions corresponding to the fiber distribution illustrated in Figure 3.6. Source: Yang, S., Tewari, A. and Gokhale, A.M., 1997. Modeling of non-uniform spatial arrangement of fibers in a ceramic matrix composite. *Acta Materialia*, 45(7), 3059–3069.

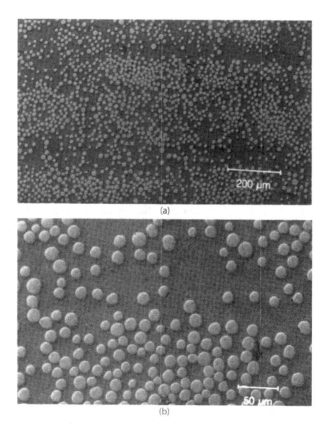

(a)

(b)

FIGURE 3.6   A micrograph of a UD composite cross section. The lower image shows a local region of focus within the larger image. Source: Yang, S., Tewari, A. and Gokhale, A.M., 1997. Modeling of non-uniform spatial arrangement of fibers in a ceramic matrix composite. *Acta Materialia*, *45*(7), 3059–3069.

of failure under a specified loading mode. In the next chapter (Chapter 4), we shall treat the concept of representation of a heterogeneous microstructure more generally, and in the chapter on failure modeling (Chapter 8) we shall employ the microstructural representation to simulation of the microstructure for the purpose of analyzing failure. Here, we note that the method described above for discriminating between different non-uniform fiber distributions in a cross section of a manufactured composite using the K-function is a good starting point in conducting studies of how the features of non-uniform fiber distributions affect failure of the composite. The failure initiation and its progression depend on the loading mode and constituent properties for a given pattern of fiber distribution. For instance, a study conducted by Pyrz [4] on different fiber distribution patterns in a composite subjected to transverse tension suggested that a more clustered fiber pattern generated higher radial tensile stress on the fiber surfaces. Much more has been done since that study and the findings of the more recent studies will be reviewed in Chapter 8 where failure analysis is treated. At this point it is of interest also to note that the scale of disorder given by $r = r_0$, as suggested by Pyrz [4], is a good starting point for examining the size of the field of observation that would be adequate for representation of the non-uniform fiber distribution.

3.3 MISALIGNMENT AND WAVINESS OF FIBERS

In most cases, except when fibers are to be placed in random orientations, a manufacturing process for composites intends to have fibers aligned in specified directions or have a specified architecture. Depending on the manufacturing process, fibers in the resulting composite deviate by different degrees from those specifications. The effects of those deviations on the composite performance have been found to be severe, especially when compression in the fibers is involved. The failure processes affected by the manufacturing defects will be treated in a later chapter. Here, we shall describe how the fiber misalignment and waviness are measured and quantified.

The observation and measurement techniques for fiber misalignment and waviness have seen continuous refinement as the microscopy and computational capabilities have increased. An early work addressed the problem of measuring fiber misalignment of small angles (<10°) [8]. The method developed is based on the observation that a circular cylindrical fiber when cut along a sectioning plane inclined at an angle to the fiber axis will display an ellipse in the cut section, as depicted in Figure 3.7. As indicated in the figure, the ellipse thus formed will have its major axis b and it will relate to the fiber diameter d depending on the inclination θ of the cut section with respect to the fiber axis. That relationship is simply, $\sin \theta = d/b$.

A manufacturing process tends to cause misalignments of fibers from an intended orientation such that the angle θ will vary for a planar array of fibers for a selected sectioning plane orientation. In [8], it was assumed that the mean of that angle measured over a region would approximate the intended fiber orientation and the

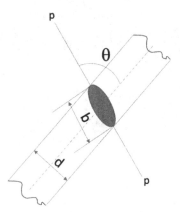

FIGURE 3.7 The elliptical cut section of a cylindrical fiber of diameter d cut by a sectioning plane p-p inclined at an angle θ to the fiber axis. The major axis of the ellipse is b.

deviation from that mean by a given fiber segment would give the misalignment angle for that fiber. The method was demonstrated for in-plane and out-of-plane fiber misalignments for a UD prepreg and for UD and 0/90 laminates [9].

While the method proposed in [8, 9] is simple, its implementation was found to be cumbersome for quantifying fiber misalignment over relatively large regions. Another method to partly improve upon this method was devised [10] where relatively low-resolution micrographs of sections parallel to the nominal fiber direction were used. Figure 3.8 shows two micrographs of a pultruded composite, one of a section cut at 6° to the pultrusion direction (Figure 3.8(a)) for use in the earlier method [8, 9] and the other along the pultrusion direction (Figure 3.8(b)). The method, proposed in [10], called multiple field image analysis (MFIA), used low-resolution images such as the one in Figure 3.8(b) along the pultrusion direction. Such images were used in a digital image-processing algorithm, which takes a pixel array over a

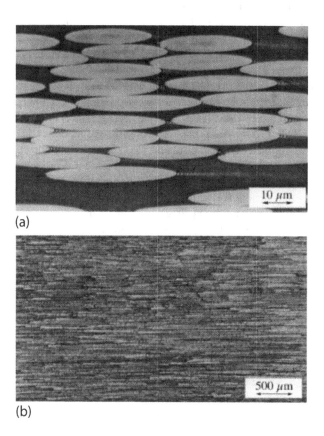

(a)

(b)

FIGURE 3.8   Images of a pultruded composite sectioned at (a) 6° to the nominal fiber direction and (b) along the nominal fiber direction. Source: Creighton, C.J., Sutcliffe, M.P.F. and Clyne, T.W., 2001. A multiple field image analysis procedure for characterisation of fibre alignment in composites. *Composites Part A: Applied Science and Manufacturing, 32*(2), 221–229.

specified length to estimate the fiber orientation. This method compared well with the fiber sectioning method [8, 9] for small regions.

While the MFIA method [10] was found to be more efficient than the earlier method of sectioning a composite at an off-axis angle [8, 9] for relatively large domains, it was still found to be slow as it required several minutes in computation time to process the collected image analysis data. The method could be made faster by utilizing two-dimensional (2D) fast Fourier transform (FFT) [11]. Thus, in this method called Fourier transform misalignment analysis (FTMA) converting the planar data into frequency domain required larger computer memory but in turn, reduced the computation time to provide the misalignment angles.

In many manufacturing methods, the fiber misalignment results from fibers or bundles of fibers becoming wavy. Two examples of this are in the micrographs in Figure 3.9 where through-thickness sections of two non-crimp fiber (NCF) composites are shown [12, 13]. This type of waviness can result in large misalignment angles. The MFIA and FTMA methods become less accurate in such cases. An alternative method called high-resolution misalignment analysis (HRMA) was proposed [12]. In this method, individual fiber segments are traced in high-resolution micrographs of cut sections obtained by optical microscopy at 50× magnification. The digitized micrographs are divided into smaller regions in which curved (wavy) fibers appear approximately straight. Many fibers (typically 50,000) are traced and their deviations from the nominal fiber direction are recorded. Figure 3.10 shows two examples of the frequencies of fiber misalignment angles for the laminates (B3 and C2) for which typical micrographs are shown in Figure 3.9. A normal probability distribution is fitted for each case, as depicted in Figure 3.10. This allows the assessment of the randomness in fiber misalignment. Thus, as seen in Figure 3.10, the laminate C2 has relatively high randomness, while the laminate B3 does not fit the normal distribution well suggesting that there is more uniformity in the fiber waviness of this laminate.

The methods discussed here so far have been based on optical micrographs as taking such micrographs, digitizing them, and using a computer are all within the reach of most laboratories and industries. It is, however, becoming increasingly possible to access X-ray computed tomography (XCT) facilities. XCT techniques allow imaging of very fine details of the composite in 3D. A study [14] used a sub-micron

a) B3

b) C2

FIGURE 3.9 Two typical micrographs showing out-of-plane waviness in two UD non-crimp fiber (NCF) composites (B3 and C2) in the through-thickness (1–3) planes, where the 1-direction is along fibers and the 3-direction is along the thickness of the composite. Source: Wilhelmsson, D., Gutkin, R., Edgren, F. and Asp, L.E., 2018. An experimental study of fibre waviness and its effects on compressive properties of unidirectional NCF composites. *Composites Part A: Applied Science and Manufacturing, 107,* 665–674.

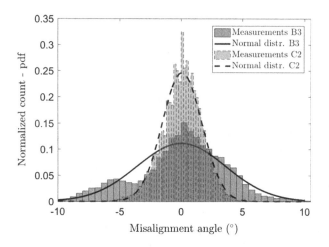

FIGURE 3.10 Measured frequencies of fiber misalignment angles in two NCF laminates B3 and C2. The fitted curves are normal distributions. Source: Wilhelmsson, D., Gutkin, R., Edgren, F. and Asp, L.E., 2018. An experimental study of fibre waviness and its effects on compressive properties of unidirectional NCF composites. *Composites Part A: Applied Science and Manufacturing, 107*, 665–674.

resolution XCT microscope to obtain 3D images of a UD composite volume. A segmentation algorithm was then used to convert the tomographic data into a numerical representation. These data were used to calculate fiber segment misorientations. The technique could sample very small volumes containing approximately 500 fibers. This limitation was mitigated in another technique [15] that could be applied to samples typically used for compression testing. This technique uses parallel slices of the examined volume in which the fiber centers are identified. These centers are then connected to determine the fiber segment directions.

The XCT technique was combined with digital image correlation (DIC) in [16]. The tomographic images were subjected to DIC where the brightness patterns resulting from the radiodensity difference between fibers and matrix were obtained on successive slices in the thickness direction. By comparing those patterns, the displacements in the patterns allowed for computing the fiber orientations. The method was useful in providing averages and standard deviations of the fiber orientations, but misalignments of individual fibers could not be determined accurately.

Another study [17] improved the accuracy of fiber misalignments by connecting with straight lines detectable fibers in the XCT images of two parallel slices in the thickness direction of the composite. The procedure used in [17] is depicted in Figure 3.11. As indicated in the figure, the volume of interest (VOI) is scanned by XCT in several slices normal to the nominal axial direction of a unidirectional composite. DIC is applied to these slices in which the fiber locations are recorded. Figure 3.11 (b) illustrates how the orientation of a fiber segment is determined by using the in-plane and out-of-plane displacements of the straight line joining the fiber positions in the top and bottom slices.

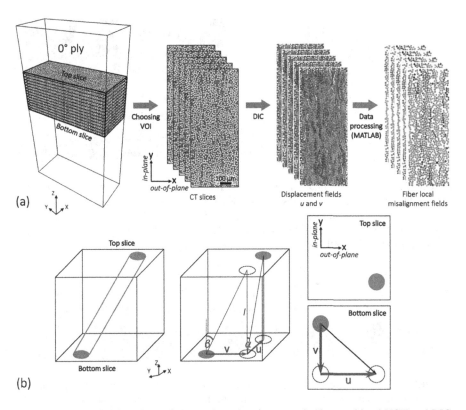

FIGURE 3.11 (a): The schematic procedure showing steps in the combined XCT and DIC technique and (b) the procedure for determining the fiber segment orientation by using the relative in-plane and out-of-plane displacements of the fiber ends. Source: Mehdikhani, M., Breite, C., Swolfs, Y., Wevers, M., Lomov, S.V. and Gorbatikh, L., 2021. Combining digital image correlation with X-ray computed tomography for characterization of fiber orientation in unidirectional composites. *Composites Part A: Applied Science and Manufacturing, 142,* 106234.

It is important to note that the XCT-based techniques can reconstruct misalignments of small fiber segments and can generate statistics of the misalignment angles. As noted above, the fiber misalignment in many cases results from initially straight fibers becoming wavy during the manufacturing process. A review of the formation of fiber waviness [18] identifies four sources of fiber waviness: Temperature gradient, consolidation, tool interaction, and fiber properties mismatch. Examples of in-plane waviness and out-of-plane waviness are shown in Figure 3.12. Such waviness occurs over a region by coordinated displacements of fibers. Within such a region the fiber misalignment angles can vary from fiber to fiber. The XCT techniques are likely not able to reconstruct the fiber waviness pattern in the entire affected region as they can image much smaller volumes. Therefore, for waviness characterization, an approach is needed that has a lower resolution than what XCT techniques typically have, such as in optical microscopy-based techniques described above. Such a technique was

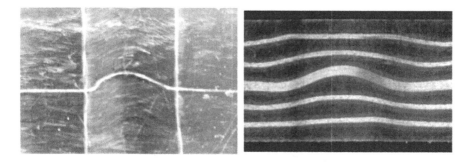

FIGURE 3.12 Examples of an in-plane fiber waviness (left) and an out-of-plane fiber waviness (right) observed in composites. Source: Kulkarni, P., Mali, K.D. and Singh, S., 2020. An overview of the formation of fibre waviness and its effect on the mechanical performance of fibre reinforced polymer composites. *Composites Part A: Applied Science and Manufacturing, 137,* 106013.

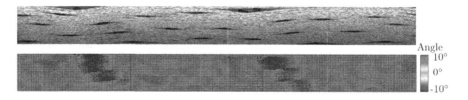

FIGURE 3.13 An optical micrograph of a section of an NCF composite (top) and the corresponding image of the section showing the fiber misalignment distribution. Source: Wilhelmsson, D., Talreja, R., Gutkin, R. and Asp, L.E., 2019. Compressive strength assessment of fibre composites based on a defect severity model. *Composites Science and Technology, 181,* 107685.

proposed in [19]. It regards a local region of fiber waviness in a plane as an ellipse and characterizes the size and shape of such ellipses by the minor axis length and the aspect ratio, respectively. Figure 3.13 shows a section of an NCF composite and the corresponding fiber misalignment distribution in the same section [19]. As can be seen, the planar regions of fiber misalignment appear as patches that can be approximated by ellipses.

3.4 MATRIX VOIDS

The conventional methods for quantification of void content in composite materials used the Archimedes principle to calculate the actual density of a sample. Then, by using the theoretical density of the sample knowing the densities of fibers and matrix, and the fiber weight fraction, the void content could be calculated. This method was not found reliable due to the uncertainties in values needed for the theoretical density computation. The fiber density can be calculated by burning off the matrix and weighing the fibers and calculating it from the known weight of the sample before the matrix burn-off and the fiber weight fraction. The inaccuracy in this method can come from any oxidation of the fibers during the matrix burn-off. An alternative to matrix burn-off is to chemically degrade the polymer.

To quantify the void shape and size in addition to the void content, optical and electron microscopy are widely used. This requires sectioning of the composite, polishing the cut surfaces, and from optical images estimating the void parameters. Three-dimensional distributions of the void parameters are then calculated from measurements on multiple sections. Figure 3.14 shows two images, one taken with an optical microscope and the other with SEM [20]. The clarity of observing the void is obviously improved in the SEM image. The SEM images can be taken sequentially of a larger area and meshed to produce a single image such as that displayed in Figure 3.15 [20] where 20 images taken at 100× magnification have been put together to show a composite cross section of 15×5 mm dimensions. Such images allow a convenient way to estimate the void parameters.

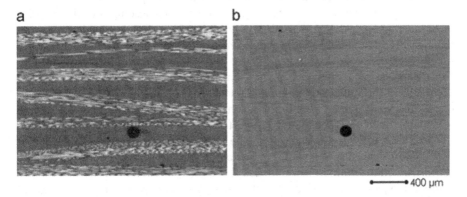

FIGURE 3.14 Optical microscopy image (a) and SEM image (b) taken of the same area of a twill weave carbon/epoxy composite. Source: Little, J.E., Yuan, X. and Jones, M.I., 2012. Characterisation of voids in fibre reinforced composite materials. *Ndt & E International*, 46, 122–127.

FIGURE 3.15 An SEM image of a composite cross section showing voids. Source: Little, J.E., Yuan, X. and Jones, M.I., 2012. Characterisation of voids in fibre reinforced composite materials. *NDT & E International*, 46, 122–127.

The void content information is useful for estimating in-plane elastic properties, as will be discussed in a later chapter. For failure analysis, however, void shape, size and, in some cases, location must also be known. For voids that are found in composites manufactured by resin transfer molding (RTM), optical microscopy was used for the quantification of voids [21]. For a disc-shaped composite, sectioning of two adjacent samples was done along a radial line in the composite. One sample was analyzed in through-the-thickness direction while the other sample was used to examine the planar void distribution. A large number (960) of images were used to perform the void parameter statistics. Although this study demonstrated that optical microscopy techniques can be used for a three-dimensional characterization of the voids, it also showed the manual work involved in performing the analysis and the uncertainties in getting robust statistics.

The limitations in the optical microscopy methods for void characterization can be reduced by resorting to X-ray micro-tomography techniques. In recent years, the increasing availability of XCT facilities has provided ways to accurately image the voids in selected (small) volumes of a composite. In an XCT technique, an X-ray beam is fired through a sample at different angles and from the measured attenuations in the scans images of a thin slice are mathematically reconstructed. By stacking such images, a volume of the composite is imaged. From contrast measurements in the images, fibers, matrix, and voids can be separated in the images. An example taken from [22] (Figure 3.16) shows how clearly the voids can be seen in the reconstructed volume. The spherical void of ~0.25 mm diameter is seen in the left part of the picture and larger voids of twice the size are visible in the right part. The clarity of voids in such images makes it possible to measure the void size and aspect ratio but many volumes must be reconstructed to obtain robust statistics of the void parameters. A summary of the information gathered by different techniques on the void parameters can be found in [23]. The trend in the research community is in the direction of using XCT techniques, and as more laboratories acquire the necessary equipment, further progress in the field is expected.

3.5   RESIN-RICH AREAS

Resin-rich areas, also described as resin-rich zones or resin pockets, are common in the practical manufacturing of composite materials. These areas are not necessarily captured in simulations of non-uniform fiber distributions if they tend to form in areas where fibers have been excessively separated or between layers of UD composites such as what is illustrated in Figure 3.17 [24]. Although such areas are easily seen by visual inspection of such optical images, their quantification is a significant challenge. A recent work [24] has proposed a methodology to automatically identify resin-rich areas in processed micrographs and to calculate statistics (histograms) of the areas. The essential steps involved in the methodology are (1) to input a microscope image, from which (2) produce a binarized image separating fibers and matrix, and (3) using the fiber

FIGURE 3.16 A 3D model of a S-glass epoxy UD composite created from XCT images. The fiber tows are in lighter grey color and the voids are in darker grey color. The matrix has been rendered invisible. The tows are approximately 1.6 mm wide and 0.5 mm high. Each tow contains ~6000 fibers. Source: Schilling, P.J., Karedla, B.R., Tatiparthi, A.K., Verges, M.A. and Herrington, P.D., 2005. X-ray computed microtomography of internal damage in fiber reinforced polymer matrix composites. *Composites Science and Technology*, *65*(14), 2071–2078.

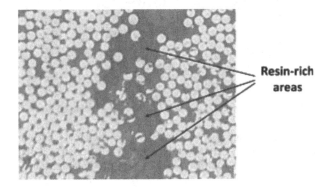

FIGURE 3.17 An optical microscopy image of a cross section of a UD composite showing the presence of resin-rich areas. Source: Li, X., Shonkwiler, S. and McMains, S., 2021. Detection of resin-rich areas for statistical analysis of fiber-reinforced polymer composites. *Composites Part B: Engineering*, *225*, 109252.

pixels as a background and matrix pixels as a foreground to generate a distance transform of the binary image using an algorithm described in [25]. An example of a micrograph and the corresponding resin-rich area imaged by the method is shown in Figure 3.18 [24].

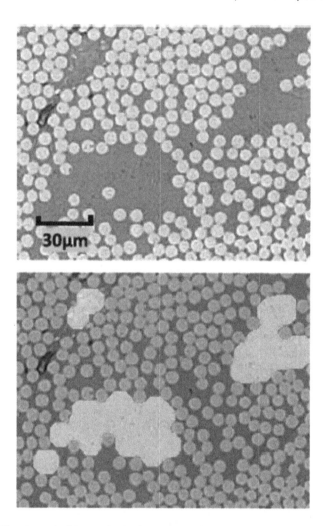

FIGURE 3.18 An optical image (top) of a cross-sectional region of UD composites and the corresponding image (bottom) generated by the algorithm described in [24]. Source: Li, X., Shonkwiler, S. and McMains, S., 2021. Detection of resin-rich areas for statistical analysis of fiber-reinforced polymer composites. *Composites Part B: Engineering*, 225, 109252.

3.6 CONCLUSION

Having described the manufacturing methods and the observed manufacturing defects in Chapter 2, this chapter has addressed the methods for quantifying those defects. The defects have been categorized as those related to fibers, matrix, and interfaces. The fiber-related defects are mainly the non-uniformity of fiber distributions and misaligned and wavy fibers. The polymer matrix can have partially cured regions, which are difficult to observe, but depending on the processing technique, the polymer matrix within the composite volume can occur as resin-rich zones. The quantification methods for non-uniform distribution of fibers depend on the images obtained by microscopy and XCT. Similarly, for misaligned and wavy fibers the quantification methods rely on the details in the microscopy images. The resin-rich areas have also been quantified from optical microscopy images. Observations of interfacial defects are difficult to make, and it is therefore common to assume those in modeling and simulation studies.

REFERENCES

1. Grufman, C. and Ellyin, F., 2007. Determining a representative volume element capturing the morphology of fibre reinforced polymer composites. *Composites Science and Technology*, 67(3–4), pp. 766–775.
2. Torquato, S. and Haslach Jr, H.W., 2002. Random heterogeneous materials: Microstructure and macroscopic properties. *Applied Mechanics Reviews*, 55(4), pp. B62–B63.
3. Pyrz, R., 1994. Quantitative description of the microstructure of composites. Part I: Morphology of unidirectional composite systems. *Composites Science and Technology*, 50(2), pp. 197–208.
4. Pyrz, R., 1994. Correlation of microstructure variability and local stress field in two-phase materials. *Materials Science and Engineering: A*, 177(1–2), pp. 253–259.
5. Ripley, B.D., 1977. Modelling spatial patterns. *Journal of the Royal Statistical Society: Series B (Methodological)*, 39(2), pp. 172–192.
6. Green, P.J. and Sibson, R., 1978. Computing Dirichlet tessellations in the plane. *The Computer Journal*, 21(2), pp. 168–173.
7. Yang, S., Tewari, A. and Gokhale, A.M., 1997. Modeling of non-uniform spatial arrangement of fibers in a ceramic matrix composite. *Acta Materialia*, 45(7), pp. 3059–3069.
8. Yurgartis, S.W., 1987. Measurement of small angle fiber misalignments in continuous fiber composites. *Composites Science and Technology*, 30(4), pp. 279–293.
9. Yurgartis, S.W., 1995. Techniques for quantification of composite mesostructured. *Composites Science and Technology*, 53, pp. 145–154.
10. Creighton, C.J., Sutcliffe, M.P.F. and Clyne, T.W., 2001. A multiple field image analysis procedure for characterisation of fibre alignment in composites. *Composites Part A: Applied Science and Manufacturing*, 32(2), pp. 221–229.
11. Kratmann, K.K., Sutcliffe, M.P.F., Lilleheden, L.T., Pyrz, R. and Thomsen, O.T., 2009. A novel image analysis procedure for measuring fibre misalignment in unidirectional fibre composites. *Composites Science and Technology*, 69(2), pp. 228–238.
12. Wilhelmsson, D. and Asp, L.E., 2018. A high resolution method for characterisation of fibre misalignment angles in composites. *Composites Science and Technology*, 165, pp. 214–221.
13. Wilhelmsson, D., Gutkin, R., Edgren, F. and Asp, L.E., 2018. An experimental study of fibre waviness and its effects on compressive properties of unidirectional NCF composites. *Composites Part A: Applied Science and Manufacturing*, 107, pp. 665–674.
14. Czabaj, M.W., Riccio, M.L. and Whitacre, W.W., 2014. Numerical reconstruction of graphite/epoxy composite microstructure based on sub-micron resolution X-ray computed tomography. *Composites Science and Technology*, 105, pp. 174–182.
15. Emerson, M.J., Jespersen, K.M., Dahl, A.B., Conradsen, K. and Mikkelsen, L.P., 2017. Individual fibre segmentation from 3D X-ray computed tomography for characterising the fibre orientation in unidirectional composite materials. *Composites Part A: Applied Science and Manufacturing*, 97, pp. 83–92.

16. Yoshimura, A., Hosoya, R., Koyanagi, J. and Ogasawara, T., 2016. X-ray computed tomography used to measure fiber orientation in CFRP laminates. *Advanced Composite Materials*, *25*(1), pp. 19–30.

17. Mehdikhani, M., Breite, C., Swolfs, Y., Wevers, M., Lomov, S.V. and Gorbatikh, L., 2021. Combining digital image correlation with X-ray computed tomography for characterization of fiber orientation in unidirectional composites. *Composites Part A: Applied Science and Manufacturing*, *142*, 106234.

18. Kulkarni, P., Mali, K.D. and Singh, S., 2020. An overview of the formation of fibre waviness and its effect on the mechanical performance of fibre reinforced polymer composites. *Composites Part A: Applied Science and Manufacturing*, *137*, 106013.

19. Wilhelmsson, D., Talreja, R., Gutkin, R. and Asp, L.E., 2019. Compressive strength assessment of fibre composites based on a defect severity model. *Composites Science and Technology*, *181*, 107685.

20. Little, J.E., Yuan, X. and Jones, M.I., 2012. Characterisation of voids in fibre reinforced composite materials. *Ndt & E International*, *46*, pp. 122–127.

21. Hamidi, Y.K., Aktas, L. and Altan, M.C., 2005. Three-dimensional features of void morphology in resin transfer molded composites. *Composites Science and Technology*, *65*(7–8), pp. 1306–1320.

22. Schilling, P.J., Karedla, B.R., Tatiparthi, A.K., Verges, M.A. and Herrington, P.D., 2005. X-ray computed microtomography of internal damage in fiber reinforced polymer matrix composites. *Composites Science and Technology*, *65*(14), pp. 2071–2078.

23. Mehdikhani, M., Gorbatikh, L., Verpoest, I. and Lomov, S.V., 2019. Voids in fiber-reinforced polymer composites: A review of their formation, characteristics, and effects on mechanical performance. *Journal of Composite Materials*, *53*, pp. 1579–1669.

24. Li, X., Shonkwiler, S. and McMains, S., 2021. Detection of resin-rich areas for statistical analysis of fiber-reinforced polymer composites. *Composites Part B: Engineering*, *225*, 109252.

25. Maurer, C.R., Qi, R. and Raghavan, V., 2003. A linear time algorithm for computing exact Euclidean distance transforms of binary images in arbitrary dimensions. *IEEE Transactions on Pattern Analysis and Machine Intelligence*, *25*(2), pp. 265–270.

4 Microstructural Descriptors

4.1 INTRODUCTION

Real materials – whether they are metals, polymers, ceramics, or their composites – have internal structures that are inherently inhomogeneous and non-uniform. For instance, if a piece of polycrystalline metal is sectioned, polished, and observed under a microscope, the grains seen have variations in their size, shape, and orientation. Historically, what was seen under a microscope became known as a "microstructure", and as our capabilities to observe the internal structure have drastically increased, e.g., by higher resolution microscopy, and recently by XCT techniques that reconstruct images of a material volume, the term microstructure is still used to refer to the internal material structure. In multiscale modeling, to be discussed later, the microscale is the lowest scale at which a material is analyzed, and it is not necessarily what is described as microstructure.

In composite materials, what is seen at a resolution where the fibers are distinct is labeled as microstructure. Thus, even though the matrix material and the fibers within a composite each have their own microstructure, the composite microstructure refers not to those, viewing them as homogeneous. This is intentional when the concern is to examine the deformation characteristics of a composite and how it fails under different loading modes. For this, the stress fields in the fibers and matrix materials are determined by models where the properties of these materials, and in some cases, those of the interface between them, are used.

4.2 THE CONCEPT OF MICROSTRUCTURAL REPRESENTATION

The lowest-order representation of a composite microstructure is the volume fraction of fibers (or matrix). This representation is easily obtained by calculating the area occupied by fibers in a composite cross section and dividing it by the total area of the cross section. The area fraction of fibers is then converted to the volume fraction by assuming straight fibers of a constant cross section. It can easily be seen that to obtain a robust estimate of the fiber volume fraction one must observe a "sufficiently large" cross-sectional area. Thus, if one imagines the observed cross section to be too small, the volume fraction will depend on the cross-sectional area, until the area is sufficiently enlarged to have the estimate of the volume fraction become independent of that area. This assumes that the cross section is statistically homogeneous,

DOI: 10.1201/9781003225737-4

i.e., it does not have the fiber distribution pattern changing over the composite volume for which the fiber volume fraction is estimated. These considerations suggest that in constructing a microstructural representation there should be a minimum volume involved. The notion of a representative volume element (RVE) of a minimum size has thus emerged. While the fiber volume fraction is useful in estimating certain simple properties such as the longitudinal modulus of elasticity, it is far from enough in estimating material characteristics associated with anisotropic deformation, and certainly not failure.

A conceptual advance in defining a minimum RVE size came from Hill's work on the effective properties of reinforced solids [1]. Hill defined the RVE to be such that it "contains sufficient number of inclusions for the apparent overall moduli to be effectively independent of the surface values of tractions and displacements". Accordingly, the minimum RVE size can be found by subjecting uniform tractions (or displacements) to the bounding surfaces of a small composite volume and increasing its size until the displacements (or tractions) on the surfaces become uniform. This notion of the RVE size is, however, applicable to the elastic properties of a composite. While it goes beyond the minimum RVE size based on the fiber volume fraction, it still does not address the local stress concentrations and stress gradients responsible for failure in composites.

Another notion of minimum RVE size emerged from considerations of two-point correlations that addressed the inter-fiber distances in composite cross sections. As discussed in Chapter 3, the intensity of inter-fiber distances given by the $g(r)$ function shows variations against the radial distance r measured from a fiber center for small values of r but tends to $g(r) = 1$ at a certain value that characterizes the **scale of disorder** [2]. This scale can be viewed as a measure of the minimum RVE size beyond which no further useful representation of inter-fiber distances is carried in the RVE. Since inter-fiber distances govern the local stresses in the matrix and on the fiber surface, the scale of disorder is a relevant parameter for failure analysis.

The scale of disorder approach to determining RVE size for observed microstructures of composites was studied on composites manufactured by different processes [3]. In that study, micrographs of cross sections were taken from composites manufactured by using prepregs as flat panels, tubes, and filament-wound tubes. In each case, the nearest neighbor fiber-center distances were measured for progressively larger square windows of observation and their frequencies were plotted against the window size. In all cases the frequencies converged towards the global distributions, confirming the existence of the scale of disorder.

The RVE size is not unique for a given composite manufactured by a given process but depends on what features of the microstructure need to be represented. Even for deformational response, e.g., elastic, or inelastic behavior, the RVE size will be different. For failure response, which is our primary focus, the RVE size depends on what local conditions (stress concentrations and stress gradients) govern failure initiation and its progression and criticality. These aspects will be treated in subsequent chapters where the modifications to RVE size warranted by the nature of the local stress fields will be discussed in detail.

4.3 SIMULATION (STATISTICAL RECONSTRUCTION) OF MICROSTRUCTURES

Once the purpose of the microstructural representation has been determined, the minimum RVE size can be decided upon, and its content can be constructed from microscopy observations of appropriate sections of a manufactured composite. In some cases, if XCT is feasible, a volume can be reconstructed, although typically such volumes are not large enough to qualify as RVEs. In any case, the RVEs can in principle be directly used to perform a numerical stress analysis, e.g., by finite element models, to determine local stress fields within the RVEs. For design purposes, however, and to gain insight into the governing processes for deformation and failure of composites, the RVEs can be subjected to parametric variations. As an example, for the same fiber volume fraction averaged over an RVE, the fiber patterns within the RVE can be varied to study the consequences of fiber clusters and resin-rich zones on the local stresses and, consequently, failure initiation. To accomplish this, relevant descriptors of fiber patterns need to be developed first, and then using these, statistically equivalent RVEs can be generated. In other words, for a given set of microstructural descriptors, multiple realizations of the fiber patterns within the RVEs can be created such that these patterns in each realization have the same RVE-averaged descriptors. In the following, some procedures for generating statistically equivalent RVEs, known as microstructure simulation, will be described.

4.3.1 SIMULATION BASED ON INTER-FIBER STATISTICAL DESCRIPTORS

As discussed in Chapter 3, the second-order intensity function $K(r)$ and its derivative $g(r)$, called the radial distribution function, form good descriptors of the inter-fiber distance variation in a composite cross section. While the $K(r)$ function provides information on the distribution of fiber centroids in the cross section, the $g(r)$ function indicates the scale of disorder [2]. Using these two functions the study [4] developed a procedure for RVE construction that will be described next.

The procedure in [4] starts by assuming that the RVE size for failure analysis would lie in the vicinity of the scale of disorder r_0. For illustration, the $K(r)$ and $g(r)$ functions obtained from SEM micrographs of a ceramic matrix composite are taken (see Figure 3.5, Chapter 3). Using $r_0 = 45$ μm from that data, a circle of this radius centered at an assumed location of a fiber centroid is drawn. Then, smaller concentric circles with equal increments are drawn (Figure 4.1a). The total number of fibers given by the composite's fiber volume fraction is then distributed within the outermost circle using the $g(r)$ function, which provides the fraction of the total number of fibers at a given value of the radial distance r. This is done by randomly distributing the number of fibers within the ring formed by circles of radius r and $r + dr$, where dr is the radial increment to the next circle. This way, multiple realizations of the RVE are generated each time using random locations within the rings for placing fibers. An example of an RVE realization is shown in Figure 4.1b, which also shows the square area enclosing the circular RVE for the convenience of applying boundary conditions to the RVE in a rectangular coordinate system.

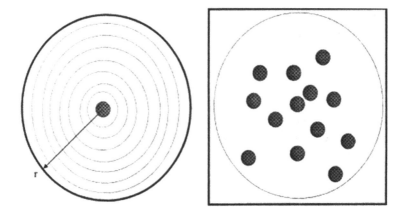

FIGURE 4.1   a. Concentric circles of increasing radius up to the outermost circle of radius equal to the scale of disorder r_0, and b. A typical realization of circular RVE enclosed within a square.

In a study of inter-fiber statistical descriptors [5] where many planar fiber distribution patterns were generated by a computer, it was found that the $K(r)$ and $g(r)$ functions were effective in distinguishing between patterns when the fiber volume fraction was relatively low (< 0.3). However, the patterns generated covered many possibilities that may not be found in real composites. Still, to be sure, it is recommended that the procedure described here be investigated for the composites under consideration to check its applicability to high fiber volume fractions.

4.3.2 SIMULATION BASED ON VORONOI CELLS

The relevance of the Voronoi cells to polycrystalline metals was obvious because of the grain size, shape, and distribution in these materials and their effects on the material properties and performance. Based on the argument that such a cell also represents a "zone of influence" of a fiber attributable to a fiber in developing the local stress field, the application of Voronoi cells to fiber-reinforced composite materials developed (see Chapter 3). One study [3] used Voronoi polygons along with the $K(r)$ and $g(r)$ functions to explore simulating the fiber patterns in composites manufactured by different processes. It was found that statistics of Voronoi cell size could also be used to distinguish between microstructures resulting from different manufacturing processes. Also, the spatial distribution of the Voronoi cell size for increasing window size converged to a global distribution. These observations encouraged the use of the Voronoi cells for simulating microstructures. However, it was found necessary to account for the fiber diameter variation, if it was large, in generating statistics for the Voronoi cell size. It is important to note that the RVE size estimated based on the Voronoi cell size is not the same as the scale of disorder r_0.

The microstructural simulation based on Voronoi size was performed specifically for the purpose of conducting stress analysis by Ghosh and co-workers [5, 6].

These studies developed a Voronoi cell finite element method (VCFEM) in which the simulated microstructure is used directly for applying surface boundary conditions and regarding the Voronoi cells as finite elements. Realizations of the microstructure were generated by using statistics of local area fraction along with the $K(r)$ and $g(r)$ functions. In the VCFEM each Voronoi cell with an embedded heterogeneity is treated as a finite element and the cell is not discretized further.

4.3.3 SIMULATION BASED ON RANDOM MICROSTRUCTURE

It is a common perception that the composite microstructure such as that viewed on a micrograph of a real composite cross section is random. It is perhaps more accurate to describe such a microstructure as non-deterministic, in the sense that the precise features such as the fiber size, shape, and the positions of the fiber centroids cannot be determined by a cause–effect type analysis. Pyrz defined the fiber centroid positions, i.e., the spatial (planar) distribution of fiber centroids, as *completely random* if any point in the micrograph of a cross section is equally likely to be occupied by all fibers and if all points in the cross sections are independent of one another [7]. Such a random process is known as a Poisson point process. It cannot be realized in real composites not only because of finite fiber size but also because a practical manufacturing process is constrained by the initial placement of fibers before resin infusion in a composite part or in a prepreg and the process parameters are controlled to some extent. In any case, the Poisson point process serves to generate a baseline spatial distribution of fiber centroids in a composite cross section. Such a pattern can be taken as a reference pattern against which patterns in real composite cross sections can be assessed. A regular pattern, e.g., a square or a hexagonal pattern, can also be used as a reference pattern.

4.3.3.1 Random Sequential Adsorption (RSA)

Since the actual fiber distribution patterns in cross sections of real (manufactured) composites are neither uniform nor Poisson point process patterns, efforts have been made to generate random fiber distribution patterns by procedures that involve finite fiber diameters and some pre-specified randomness. One commonly used procedure for generating a random fiber pattern in a two-dimensional region is an extension of the Poisson point process where one starts with a square region of a certain size and determines how many fibers of a given diameter must be placed in the region to have the needed fiber volume (area) fraction. This number of fibers then must be distributed randomly within the region. In [8] the author considered the problem of randomly distributing circular disks in a square plane and developed an algorithm that is known as random sequential adsorption (RSA). As the name suggests, one starts by randomly placing a disk (fiber) in the square plane (a cross section) and sequentially fills the plane with the remaining disks using a random number generator until the required number of fibers has been placed while avoiding overlaps with already filled disks. It should be realized that the fiber distribution patterns generated in this manner, called hard-core patterns, do not in any way simulate a manufacturing process for composite materials as such a process does not proceed following the RSA procedure. It is nevertheless a convenient way for generating random fiber distribution patterns that allow conducting parametric studies by varying the fiber diameter and fiber volume fraction. One can still use an appropriate criterion for estimating the minimum size of the RVE generated this way, such as the scale of disorder given by the $g(r)$ function as discussed previously. This does not assure representation of the actual manufacturing process, only that further increasing the size of the square plane does not add statistically significant disorder.

4.3.3.2   Perturbation (Shaking) Methods

The RSA procedure was found to be limited by the so-called jamming limit, i.e., the maximum coverage measured in fiber area fraction that can be accomplished without fiber overlap. For typical diameters of fibers in polymer matrix composites this limit falls below the maximum fiber volume fraction that can theoretically be achieved by maximum packing and often below the fiber volume fraction in high-performance composites. Efforts have been made to add more fibers by conducting random fiber displacements, known as "shaking", after a hard-core fiber pattern has been generated by the RSA procedure. In [9], for instance, each disk (fiber) is given a small random displacement independent of its neighbors' positions to release the fibers from their jamming configuration. An algorithm was developed in [10] to iteratively displace fibers to create more unfilled areas for accommodating additional fibers. Additionally, fibers in the areas in the outskirts at the boundaries of the window were given appropriate displacements to increase the fiber volume fraction.

A different procedure than the RSA was devised in [11, 12] to avoid the jamming configuration. In [11] a first disk was used at a random location and its neighboring fibers were placed using pre-specified distributions of distances to the first and second nearest neighbors. The filling procedure was continued sequentially until the square window was filled with the required number of disks. In [12] a similar procedure called random sequential expansion (RSE) was used where minimum and maximum inter-fiber distances were used in the sequential filling process. The procedure is sensitive to the choice of the minimum and maximum inter-fiber distances and needs additional intervention to avoid underfilling or overfilling the window with disks.

Rather than impose shaking on a hard-core pattern one can perturb a uniform pattern in a random manner to generate a random pattern. The advantage here is that the fiber volume fraction is not a limitation as it is in the jamming limit inherent in the RSA procedure. In [13] the authors started with a uniform square pattern of fiber centroids in a dilute dispersion with different fiber diameters and by displacing the fibers randomly in two orthogonal directions they generated a random fiber distribution pattern with higher fiber volume fraction. The idea of starting with a uniform fiber pattern was pursued further in [14, 15]. In [14] a preselected number of fibers given by the fiber diameter and fiber volume fraction were placed in a square pattern and the fibers were then randomly and sequentially perturbed until the minimum specified distance between the neighboring fiber surfaces was achieved. It took on the order of 1000 steps (iterations) to achieve the final fiber configuration. Figure 4.2 illustrates the final configurations for four different cases of minimum fiber surface

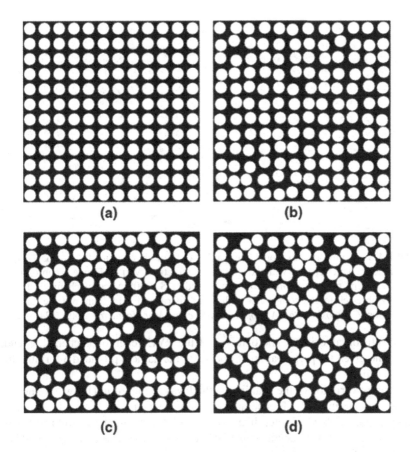

FIGURE 4.2 Final perturbed fiber patterns resulting from random perturbations of fibers of 50% volume fraction in a square pattern with specified minimum fiber surface distance of (a) 20%, (b) 15%, (c) 10%, and (d) 5% of fiber diameter. Source: Chen, X. and Papathanasiou, T.D., 2004. Interface stress distributions in transversely loaded continuous fiber composites: parallel computation in multi-fiber RVEs using the boundary element method. *Composites Science and Technology, 64*(9), 1101–1114.

separation distances given as percentages of the fiber diameter. As seen in Figure 4.2, increasing fiber clustering is achieved by reducing the minimum fiber surface separation. In [15] a different scheme for the shaking process for generating random fiber distribution patterns was proposed (Figure 4.3). Taking a hexagonal pattern as the initial pattern, fibers were given random perturbation in radial and angular directions, iteratively, until two reference frames, one outer frame (Frame 1) and an inner frame (Frame 2) reached approximately the same specified (initial) fiber volume fraction. Figure 4.3 displays one example where the initial fiber volume fraction is 65%. An interesting procedure for shaking was proposed in [16] where starting with a square or hexagonal pattern each fiber was given a unit velocity in a random direction and if any two fibers were found to overlap, their velocities were changed by an

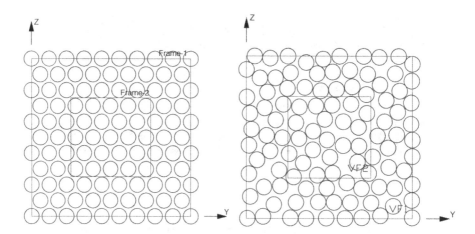

FIGURE 4.3 A uniform hexagonal fiber pattern (left) of fiber volume fraction 65%, and the shaken random pattern (right) with the fiber volume fractions VF1 (in Frame 1) approximately equal to VF2 (in Frame 2). Source: Wongsto, A. and Li, S., 2005. Micromechanical FE analysis of UD fibre-reinforced composites with fibres distributed at random over the transverse cross-section. *Composites Part A: Applied Science and Manufacturing, 36*(9), 1246–1266.

elastic collision algorithm. The outcome of the RVE generation scheme depended on the time given for perturbation of fibers and the authors provided a formula for a workable time.

One more variation of the shaking method was proposed in [17] where the starting point was an initial RVE generated by filling a square window with fibers taken randomly from a fiber diameter distribution until the specified fiber volume fraction was reached. This RVE generated as a Poisson point process is then perturbed iteratively to satisfy a constrained optimization condition. Convergence to the optimal solution depends on how non-uniform the fiber diameter distribution is. Figure 4.4 displays one example of constant fiber diameter for which the initial RVE is shown in Figure 4.4(a) and the final fiber pattern after completion of the optimization is shown in Figure 4.4(b). Since a Poisson point process is used for the initial RVE, some fibers overlap. This overlap is removed by the shaking process that follows the prescribed optimization condition.

4.3.4 SIMULATION BASED ON FIBER MOBILITY

The simulation procedures described above focus on the randomness of the fiber spatial distribution as seen in a composite cross section transverse to fibers. Every scheme devised to generate realizations of the random fiber pattern provides a means to study the effects of the fiber distribution randomness on selected deformation and/or failure characteristics such as effective elastic properties and crack initiation. These effects are then described with respect to statistical descriptors such as two-point correlation functions that describe random variations of inter-fiber spacing. While useful insights into the effects of non-uniformity of fiber distribution are

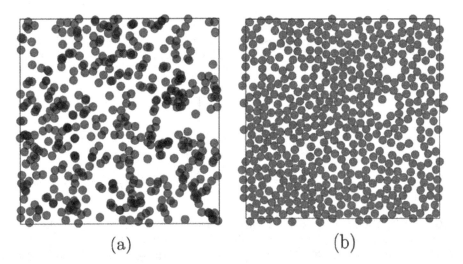

(a)   (b)

FIGURE 4.4 Initial RVE filled with fibers of one diameter (a), and the final RVE after constrained optimization (b). Source: Pathan, M.V., Tagarielli, V.L., Patsias, S. and Baiz-Villafranca, P.M., 2017. A new algorithm to generate representative volume elements of composites with cylindrical or spherical fillers. *Composites Part B: Engineering, 110*, 267–278.

gained from such studies, it is not clear how the knowledge gained can be used to conduct the cost/performance trade-off discussed in Chapter 1. For this trade-off one needs to have at the minimum some indication of how a manufacturing process induces deviations from the desired uniform microstructure. To move in this direction, studies were conducted to quantify the non-uniformity in the random microstructure in terms of fiber mobility induced during a manufacturing process [18, 19]. This type of quantification addresses defect severity more generally and will be described later for fiber misalignment as well.

4.3.4.1 Quantifying Fiber Distribution Non-uniformity

The starting point in [18] is the shaking process using a square area containing fibers in a uniform square pattern with a prescribed fiber volume fraction. The concept underlying the proposed quantification of non-uniformity is a measure based on how much a fiber can deviate from its position in a uniform pattern. To do this, the method considers the distance available between surfaces of adjacent fibers in a uniform pattern. This distance depends on the fiber diameter and the fiber volume fraction and sets the limit to the deviation of a fiber center from its position in the uniform pattern. A linear measure is adopted, where a zero degree of non-uniformity refers to the uniform pattern and a 100% degree of non-uniformity corresponds to the collective positions of fibers resulting from the maximum available deviation. To illustrate the construction of an RVE, consider four neighboring fibers in a square pattern shown in Figure 4.5. The normal distance between the two nearest fiber surfaces is denoted by "x". From the center of a given fiber (top-right in Figure 4.5),

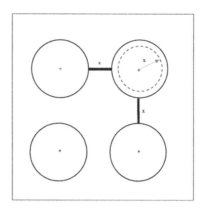

FIGURE 4.5 Using four fibers in uniformly distributed fibers, the figure shows the distance between the surfaces of the two nearest neighboring fibers, labeled x. This is the maximum available extent of a fiber-center displacement without overlapping of fibers and is taken as the circle of radius x within which the fiber center can be moved. Source: Elnekhaily, S.A. and Talreja, R., 2018. Damage initiation in unidirectional fiber composites with different degrees of nonuniform fiber distribution. *Composites Science and Technology, 155,* 22–32.

draw a circle of radius x. For the degree of non-uniformity of 100%, assume that the fiber mobility is restricted by manufacturing, i.e., without causing fiber overlap, such that the center of the considered fiber takes any random position within this circle. With the assumed linear measure, for any degree of non-uniformity less than 100%, the radius of the circle will be proportionately smaller. Thus, for a 0% degree of non-uniformity, the circle will shrink to the fiber center, giving no deviation from the uniform pattern.

The algorithm to generate a realization of RVE of a selected degree of non-uniformity is as follows [18]. Each fiber center is given a random position within the circle of radius x (described above) times the degree of non-uniformity (as a fraction). This circle is divided into concentric sub-circles with a radius ranging from zero to that of the outer circle in increments of 0.1% of the outer circle. A circle is then randomly selected from these sub-circles and the center of the displaced fiber is placed on this circle by a random selection of the angular position between zero and 2π measured from a reference axis. The algorithm then checks if there is any overlapping, i.e., if the distance between any two adjacent fiber centers is less than the fiber diameter, and if any fiber is totally or partially outside the RVE region. If so, then this fiber is given a new random position starting again from its initial uniform one. If not, then a new position is given to the next fiber, as explained above, and so on. This procedure does not change the fiber volume fraction from the initial one in the uniform distribution.

The deviation from the initial uniform square pattern resulting from the shaking process for a given degree of non-uniformity depends on the fiber volume fraction and the fiber radius. For instance, if all fiber surfaces are in contact with the surfaces of the neighboring fibers, then the distance available for fibers to move, x = 0 (see

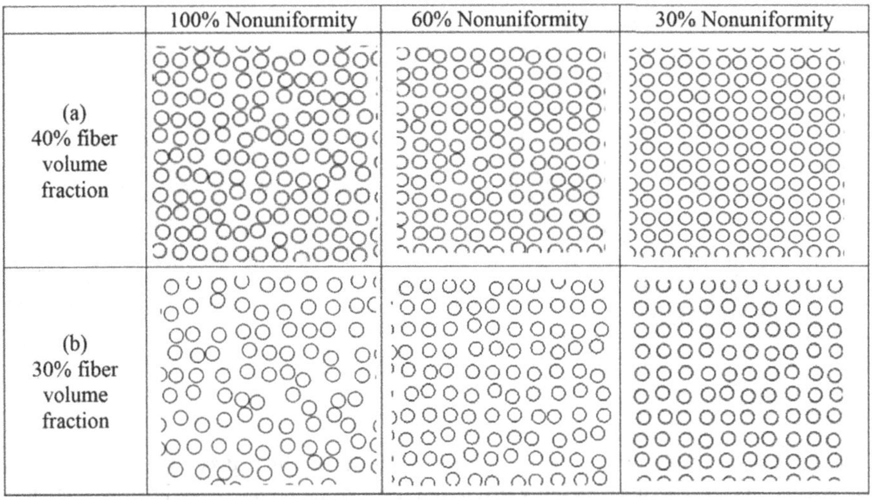

	100% Nonuniformity	60% Nonuniformity	30% Nonuniformity
(a) 40% fiber volume fraction			
(b) 30% fiber volume fraction			

FIGURE 4.6 Parts of RVE realizations for (a) 40% and (b) 30% fiber volume fraction at different degrees of non-uniformities of fiber distribution. Source: Elnekhaily, S.A. and Talreja, R., 2018. Damage initiation in unidirectional fiber composites with different degrees of nonuniform fiber distribution. *Composites Science and Technology*, *155*, 22–32.

Figure 4.5), gives no deviation. As this distance increases, e.g., by decreasing fiber volume fraction for the same fiber radius, the deviation from the initial pattern for a given degree of non-uniformity also increases. Figure 4.6 illustrates this effect by showing RVE realizations generated for two different fiber volume fractions at three different degrees of non-uniformity. Note that at 0% degree of non-uniformity, all distributions in Figure 4.6 collapse into a regular square pattern.

4.3.4.2 Quantifying Fiber Clustering

Random perturbations given to fibers in a uniform pattern simulate situations such as resin transfer molding where the fibers have been placed in a preform in a relatively cluster-free configuration and resin infusion then displaces fibers to induce non-uniformity. Another relevant case is that of a prepreg manufacturing process where fiber bundles are run through a comb to separate fibers with equal spacing and impregnation with resin then perturbs this spacing. There are other situations of manufacturing by resin infusion processes where the starting point is dry fiber bundles in which resin penetrates separating the fibers while filling the inter-fiber spaces. For such cases, another RVE generating scheme was devised in [19] and is described below.

One starts with a dry fiber bundle illustrated in Figure 4.7(a). Before the infusion of resin into the fiber bundle, the fiber configuration can be described as consisting of fiber at the center of the bundle surrounded by fibers in concentric rings of increasing radii measured from the center of the central fiber. Using a polar coordinate system, the position of the center of any fiber in the bundle is given by the coordinates (r, θ),

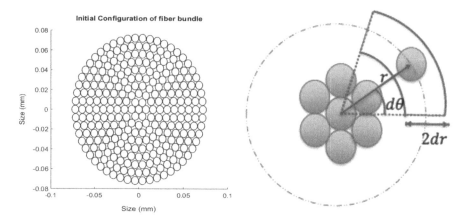

FIGURE 4.7 (a) a dry fiber bundle described by fibers (circular disks) placed in concentric circles, and (b) the fiber mobility given by ranges of radial and angular displacements. Source: Sudhir, A. and Talreja, R., 2019. Simulation of manufacturing induced fiber clustering and matrix voids and their effect on transverse crack formation in unidirectional composites. *Composites Part A: Applied Science and Manufacturing*, *127*, 105620.

see Figure 4.7(b). On resin infusion, a fiber of these coordinates displaces to another position depending on its "mobility" during the infusion process. For instance, the central fiber remains immobile, while a fiber in the outermost circle of the bundle has the most radial mobility. We specify the radial mobility as $dr = kr$, where k is a pre-specified number decided by the fiber volume fraction expected on completion of the resin infusion. The mobility of the fiber in the tangential direction is specified by the perturbation $d\theta$ of the angular coordinate θ. To capture the randomness of the resin infusion process, the total region of possible movement of a fiber, illustrated in Figure 4.7(b) for one fiber, is divided up into discrete points (r_i, θ_i), $i = 1, 2, \ldots,$ 50,000, that can be taken up by the center of the fiber, and the actual position is then decided by a random number drawn from a uniform probability distribution. However, to qualify as an acceptable position, the displaced position of the fiber must comply with the constraint that the fiber does not overlap with another fiber. Additionally, fiber surfaces are not allowed to be in contact, but a minimum distance between two neighboring fibers is kept allowing placing a minimum of two elements of the finite element mesh in the matrix between fibers.

Figure 4.8 illustrates the fiber mobility process in stages from the initial dry bundle configuration until the final configuration, which is achieved when all fibers in the smallest circle in the bundle have been displaced. It is noted that the clustering of fibers in the final configuration depends on the value of k specified in the fiber mobility process. Thus, a smaller value will lead to a higher degree of fiber clustering. Figure 4.9 shows for illustration two cases of the final configuration of fibers for $dr = 0.1r$ and $0.4r$ for the same $d\theta = 15°$. The effect of radial mobility on fiber clustering can be seen clearly.

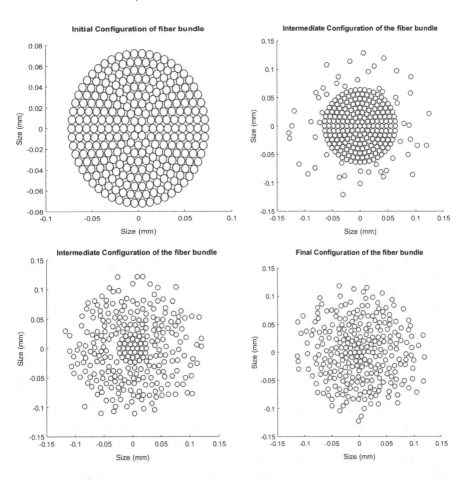

FIGURE 4.8 The spreading of fibers in the resin infusion process from the initial dry bundle (upper left) to the final configuration (lower right) with two intermediate stages shown for illustration. Source: Sudhir, A. and Talreja, R., 2019. Simulation of manufacturing induced fiber clustering and matrix voids and their effect on transverse crack formation in unidirectional composites. *Composites Part A: Applied Science and Manufacturing, 127,* 105620.

4.4 STATISTICAL EQUIVALENCY OF RVEs

As discussed previously, real composite manufacturing processes do not produce uniform fiber distributions. Also, they do not produce the so-called completely random distribution given by the Poisson point process. The many approaches described above for generating RVEs based on micrographs of actual cross sections or by devising schemes to induce randomness in initially uniform patterns or modifying randomness in hard-core patterns are based on the concept of statistical equivalency, i.e., two fiber distributions are statistically equivalent if they have the same prespecified statistics. Thus, the statistical equivalency is not absolute but depends on

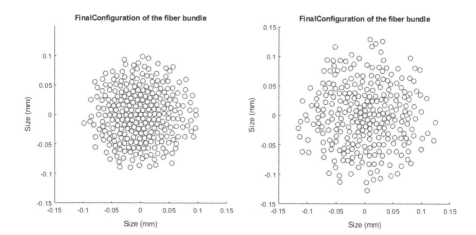

FIGURE 4.9   A RVE realization for $dr = 0.1r$ (left) and $dr = 0.4r$ (right). The angular mobility range $d\theta = 15°$ in both cases. Source: Sudhir, A. and Talreja, R., 2019. Simulation of manufacturing induced fiber clustering and matrix voids and their effect on transverse crack formation in unidirectional composites. *Composites Part A: Applied Science and Manufacturing, 127,* 105620.

the choice of the statistics. Early on in this field, it was suggested that statistics of two-point correlations such as the second-order intensity function $K(r)$ was a useful way to characterize the fiber distributions [2] based on the observation that the local stress fields in a composite were governed by the inter-fiber distances. The radial distribution function $g(r)$, derived from $K(r)$ provided the scale of disorder [2]. These two functions have been the most used bases for establishing statistical equivalency between RVEs. Other statistics related to Voronoi cells [3] and distributions of first and second nearest neighbors [17] have also been used but mostly for characterizing aspects of fiber distributions that are assumed to be of importance in producing local stress concentrations and stress gradients.

    Before statistical equivalency is examined it is important to establish whether the window chosen to place fibers in has adequate size to give robust statistics. This problem of minimum RVE size has been studied by many and has been discussed in Chapter 3 with respect to the scale of disorder that is indicated by the $g(r)$ function. It is noted here that just as statistical equivalency is not an absolute concept, so is the case with the minimum RVE size. This problem has been addressed in [20] in an extensive manner using different criteria for adequacy of the RVE size ranging from pure geometry aspects to averages of elastic properties and local stress field characteristics. With the selected criteria and applying them to a carbon/epoxy composite in a square window of side length L subjected to uniform transverse tension, a large variation in the minimum value of L was found. In terms of the fiber diameter d, the value of L varied between 10- and 50-times d. Based on inter-fiber distance distributions, L was found to be $40d$. The other values based on local stress fields and effective composite properties cannot be generalized as they depend on

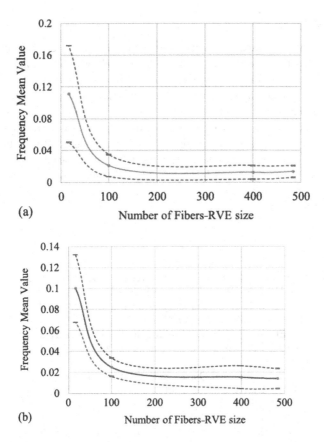

(a)

(b)

FIGURE 4.10 The maximum, mean and minimum frequencies of occurrence of the first nearest neighbor distances averaged over all distances in RVE plotted for 50% fiber volume fraction for (a) 100% and (b) 30% degrees of non-uniformity against the RVE size in number of fibers. Source: Elnekhaily, S.A. and Talreja, R., 2018. Damage initiation in unidirectional fiber composites with different degrees of nonuniform fiber distribution. *Composites Science and Technology, 155,* 22–32.

the constituent properties and fiber volume fraction. For illustration, Figure 4.10 [18] shows the variation of the first nearest neighbor distances of fibers averaged over all fibers in RVE with the RVE size measured in the number of fibers. The curves are based on multiple realizations and show the maximum, mean, and minimum values of those realizations. It can be noted from these results that a minimum RVE size can be estimated based on fiber distribution statistics and that this size depends slightly on the degree of non-uniformity of fiber distribution. It is recommended that as a first estimate, the RVE size should be based on the fiber distribution realizations. A refinement (adjustment) in the RVE size so obtained can then be based on local stress fields as they depend additionally on the constituent material properties.

4.5 CONCLUSION

This chapter has addressed developing descriptors of the observed features of the composite microstructures for the purposes of modeling initiation and progression of failure. The concept of representative volume elements as a means for describing statistically equivalent material states has been discussed. Various simulation strategies for representing these states to conduct stress and failure analysis have been described. A method for quantifying the non-uniformity of fiber distribution in the cross section of a composite that facilitates relating the degree of non-uniformity to the initiation of failure has then been described. The microstructure representation discussed here is essential to conducting the physical modeling discussed in Chapter 1 and pertains to treating the real material states, both initial and current, in a multiscale modeling effort.

REFERENCES

1. Hill, R., 1963. Elastic properties of reinforced solids: Some theoretical principles. *Journal of the Mechanics and Physics of Solids*, *11*(5), pp. 357–372.
2. Pyrz, R., 1994. Correlation of microstructure variability and local stress field in two-phase materials. *Materials Science and Engineering: A*, *177*(1–2), pp. 253–259.
3. Grufman, C., Ellyin, F., 2007. Determining a representative volume element capturing the morphology of fibre reinforced polymer composites. *Composites Science and Technology*, *67*, pp. 766–775.
4. Bulsara, V.N., Talreja, R. and Qu, J., 1999. Damage initiation under transverse loading of unidirectional composites with arbitrarily distributed fibers. *Composites Science and Technology*, *59*(5), pp. 673–682.
5. Ghosh, S., Nowak, Z. and Lee, K., 1997. Quantitative characterization and modeling of composite microstructures by Voronoi cells. *Acta Materialia*, *45*(6), pp. 2215–2234.
6. Ghosh, S., 2011. *Micromechanical Analysis and Multi-Scale Modeling Using the Voronoi Cell Finite Element Method*. Boca Raton: CRC Press.
7. Pyrz, R., 1994. Quantitative description of the microstructure of composites. Part I: Morphology of unidirectional composite systems. *Composites Science and Technology*, *50*(2), pp. 197–208.
8. Feder, J., 1980. Random sequential adsorption. *Journal of Theoretical Biology*, *87*(2), pp. 237–254.
9. Buryachenko, V.A., Pagano, N.J., Kim, R.Y. and Spowart, J.E., 2003. Quantitative description and numerical simulation of random microstructures of composites and their effective elastic moduli. *International Journal of Solids and Structures*, *40*(1), pp. 47–72.
10. Melro, A.R., Camanho, P.P. and Pinho, S.T., 2008. Generation of random distribution of fibres in long-fibre reinforced composites. *Composites Science and Technology*, *68*(9), pp. 2092–2102.
11. Vaughan, T.J. and McCarthy, C.T., 2010. A combined experimental–numerical approach for generating statistically equivalent fibre distributions for high strength laminated composite materials. *Composites Science and Technology*, *70*(2), pp. 291–297.
12. Yang, L., Yan, Y., Ran, Z. and Liu, Y., 2013. A new method for generating random fibre distributions for fibre reinforced composites. *Composites Science and Technology*, *76*, pp. 14–20.

13. Gusev, A.A., Hine, P.J. and Ward, I.M., 2000. Fiber packing and elastic properties of a transversely random unidirectional glass/epoxy composite. *Composites Science and Technology*, *60*(4), pp. 535–541.
14. Chen, X. and Papathanasiou, T.D., 2004. Interface stress distributions in transversely loaded continuous fiber composites: Parallel computation in multi-fiber RVEs using the boundary element method. *Composites Science and Technology*, *64*(9), pp. 1101–1114.
15. Wongsto, A. and Li, S., 2005. Micromechanical FE analysis of UD fibre-reinforced composites with fibres distributed at random over the transverse cross-section. *Composites Part A: Applied Science and Manufacturing*, *36*(9), pp. 1246–1266.
16. Zhang, T. and Yan, Y., 2017. A comparison between random model and periodic model for fiber-reinforced composites based on a new method for generating fiber distributions. *Polymer Composites*, *38*(1), pp. 77–86.
17. Pathan, M.V., Tagarielli, V.L., Patsias, S. and Baiz-Villafranca, P.M., 2017. A new algorithm to generate representative volume elements of composites with cylindrical or spherical fillers. *Composites Part B: Engineering*, *110*, pp. 267–278.
18. Elnekhaily, S.A. and Talreja, R., 2018. Damage initiation in unidirectional fiber composites with different degrees of nonuniform fiber distribution. *Composites Science and Technology*, *155*, pp. 22–32.
19. Sudhir, A. and Talreja, R., 2019. Simulation of manufacturing induced fiber clustering and matrix voids and their effect on transverse crack formation in unidirectional composites. *Composites Part A: Applied Science and Manufacturing*, *127*, 105620.
20. Trias, D., Costa, J., Turon, A. and Hurtado, J.E., 2006. Determination of the critical size of a statistical representative volume element (SRVE) for carbon reinforced polymers. *Acta Materialia*, *54*(13), pp. 3471–3484.

5 Failure in Unidirectional Composites in the Presence of Defects

5.1 INTRODUCTION

Failure in unidirectionally fiber-reinforced (UD) composite materials is often described as failure of matrix, fibers, and the interfaces between them. Logically, in any two-phase material, the three possible components of failure are the failure of the two phases and the interfaces where they are bonded. However, under a given loading, in which phase the failure initiates first, how it impacts other failure components, how, on further loading, the three failure components interact and progress, and, finally, what brings about the final failure. These are all questions that have different answers in different situations. What makes the failure process further complex is the effects of having manufacturing defects present within the composite volume. In the previous chapters, we discussed manufacturing processes (Chapter 2), manufacturing defects (Chapter 3), and quantification and simulation of manufacturing defects (Chapter 4). Here, we shall describe which failure mechanisms operate in UD composites under different simple loading modes (axial tension, axial compression, transverse tension, transverse compression, and in-plane shear) before in each case examining the effects of the presence of manufacture defects. In the next chapter, discussions of the failure process will be continued for laminates of straight fibers and other fiber architectures.

5.2 FAILURE UNDER LONGITUDINAL TENSION

Among the earliest studies of failure under axial tension of UD composites was the study reported in [1]. In this study, the authors tested dry bundles of carbon fibers as well as UD composites made of carbon/epoxy prepregs of 0.35 fiber volume fraction. The dry bundles were obtained by dissolving the resin from the prepregs while the composites were prepared in fully cured or semi-cured forms. The fully cured specimens were manufactured by hot pressing in a normal way while the semi-cured specimens were prepared by reducing the cure temperature in the gauge section. The tensile strength was found to show a larger scatter for fully cured composites than for semi-cured composites and fiber bundles. Based on acoustic emissions and SEM observations of failed composites the authors found that the successive failure of fibers was localized within the composite in what may

DOI: 10.1201/9781003225737-5

be described as sub-bundles and that several such failed sub-bundles connected through sheared planes between them to cause the final failure. Another study [2] independently confirmed most of the observations and inferences made in [1]. In [2] a carbon/epoxy composite of 0.65 fiber volume fraction was made by curing the prepregs in an autoclave and was tested in axial tension. SEM was performed on the failed surfaces of the specimens and the failed fibers in those surfaces were painstakingly traced. The authors inferred from those observations that small zones of failed fibers were formed independently.

Observing the tensile failure process in UD composites by optical microscopy and SEM has severe limitations. Therefore, not much progress was made in understanding the process until high-resolution XCT was employed. The first study that focused on the sequential fiber failure process in the tensile failure of UD composites was [3]. This study used 125 quartz fibers in an epoxy matrix to make a model composite that was tested in a tensile tester while XCT scans were taken at increasing stress. Figure 5.1 shows a series of X-ray radiographs where single fiber breaks (singlets), two neighboring fiber breaks (doublets), and an increasing number

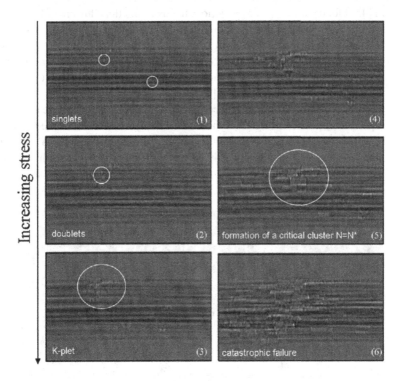

FIGURE 5.1 A series of XCT radiographs showing an increasing number of neighboring fibers breaking to form a critical cluster that produced failure. Source: Aroush, D.R.B., Maire, E., Gauthier, C., Youssef, S., Cloetens, P. and Wagner, H.D., 2006. A study of fracture of unidirectional composites using in situ high-resolution synchrotron X-ray microtomography. *Composites Science and Technology*, *66*(10), 1348–1353.

of neighboring fiber breaks (k-plets) are marked [3]. A critical cluster of fiber breaks before imminent failure is also marked.

More studies of the fiber failure process in axial tension were conducted later using XCT on carbon/epoxy $[90/0]_s$ laminates [4, 5]. In [4] the laminate was given a double notch and the notch region was CT-scanned from the edge. The observed area showed damage in the form of transverse ply cracks, 0/90 delamination, and axial splits. Focusing on the fiber breaks it was found that the sequence of fiber failures showed similar behavior to that described in [3]. The number of fiber breaks was plotted against the nominal stress in the 0-degree plies and showed an exponential increase. On separating the fiber breaks as singlets, doublets, etc., it was found that the singlets started earliest and accumulated to the largest number, while doublets, triplets, etc., started increasingly later and reached a decreasing final number. The largest k-plet recorded was for $k = 14$ [4, 5]. Figure 5.2 [4] displays CT-scans of transverse slices normal to the fibers indicating clear details of the broken fiber clusters. A similar study [6] conducted using fast XCT found that a substantial number of fiber breaks occurred very close to the final failure. That study also found specimen-to-specimen variation in fiber break accumulation. Another study [7] on the same material as in [4, 5] focused on cluster formation. It was found that clusters of fiber breaks formed in one load step but did not increase in size (number of broken fibers) as the load was increased. Instead, more clusters were formed of different size. While XCT has the potential to provide details with clarity not possible by other means, interpretation of failure events using this technique requires advanced skills and understanding of the capabilities of the technique [8].

The studies of tensile failure of UD composites reported in [4–6] have all been on carbon/epoxy composites using the synchrotron X-ray source where the CT is particularly well suited because of the low contrast in these materials [8]. The motivation of these studies has been to not only gain insight into the tensile failure process but to also assess models put forth for the prediction of the tensile strength. The

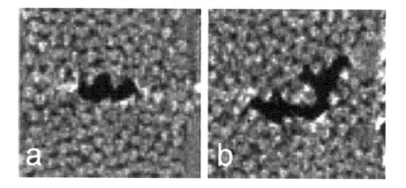

FIGURE 5.2   Planar sections taken with XCT near clusters of (a) 5 and (b) 14 broken fibers. Source: Scott, A.E., Mavrogordato, M., Wright, P., Sinclair, I. and Spearing, S.M., 2011. In situ fibre fracture measurement in carbon–epoxy laminates using high resolution computed tomography. *Composites Science and Technology*, 71(12), 1471–1477.

focus of observations has been the fiber breakage sequence and accumulation and the formation of fiber breakage clusters. Yet another study of fiber breakage clusters was reported in [9]. These authors used UD carbon/epoxy composites, not cross-ply laminates as in [4–6], in three different fiber volume fractions (38, 43, and 49%) manufactured by VARTM. XCT was performed on the specimens while in a tensile tester and stereology was used on the images to quantify the fiber breakage cluster size. The cluster size was found to increase with the fiber volume fraction.

The roles, if any, played by failure within the matrix and at the fiber/matrix interface have not been reported in the XCT observations. An attempt, however, was made to measure the axial strain along one or more broken fibers using digital volume correlation (DVC) in conjunction with XCT [10]. The same carbon epoxy $[90/0]_s$ was tested in axial tension and the XCT technique was employed as in [4–6] except this time the DVC was used to measure the volumetric displacement field around broken fibers. While the DVC technique cannot directly detect fiber/matrix debonding, the axial strain in the volume around broken fibers provided the length from the broken fiber end beyond which the strain was recovered to the pre-failure level. The data gathered suggest that as a singlet evolved into a doublet, the recovery length increased substantially.

Finding evidence of matrix cracking and fiber/matrix debonding under an axial tensile load by the XCT technique is not easy. Despite the high resolution of this technique, the scale at which these mechanisms operate has made it difficult to track their initiation and progression. It has therefore been natural to aid the observations with interpretations based on modeling and simulation studies. Most modeling studies follow the assumptions of the statistical strength theories for fibrous composite materials [11]. Such theories do not consider the growth of matrix cracks and fiber/matrix debonding in the process of tensile failure of UD composites. There are, however, indications from single-fiber composite studies that matrix cracks and fiber/matrix debonding occur in conjunction with a fiber break under tension. These cracking mechanisms depend also on the fiber/matrix interface characteristics. An example of this was reported in [12] where different surface treatments were given to glass fibers inducing different interfacial shear strengths when reinforced with an epoxy. On failure of a single fiber embedded in the matrix, the ensuing matrix cracking and fiber/matrix interface debonding depended on the interfacial strength. Two examples of the cracking configurations are displayed in Figure 5.3 where a stronger interface is seen to have longer matrix cracks and little debonding while the opposite occurs with a reduction of the interfacial strength [12]. In a composite with multiple fibers, it is expected that the two cracking modes shown in Figure 5.3 will be altered because of stress interaction with fibers neighboring the broken fiber. One modeling study has shown how the path of a growing matrix crack is affected by the presence of a fiber in the crack's way [13]. Figure 5.4 displays the trajectories of debond cracks kinked out in the matrix at different debond lengths [13]. The effect of the neighboring crack is to divert the crack's straight path increasingly towards the fiber's surface, increasing the debond length before the crack kink-out. It is expected that when the second fiber breaks, the crack emanating from the first fiber break will continue towards another (third) fiber, eventually creating a fiber breakage cluster if the local

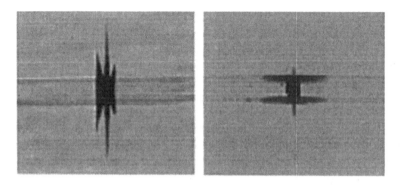

FIGURE 5.3 Occurrence of matrix cracking and fiber/matrix debonding from a fiber break under the tension of a single-fiber composite. Left: high interface strength, right: low interface strength. Source: Zhao, F.M. and Takeda, N., 2000. Effect of interfacial adhesion and statistical fiber strength on tensile strength of unidirectional glass fiber/epoxy composites. Part I: experiment results. *Composites Part A: Applied Science and Manufacturing*, *31*(11), 1203–1214.

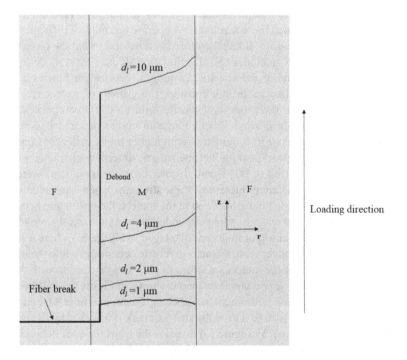

FIGURE 5.4 Trajectories of fiber/matrix debond cracks initiated at a fiber break under tension and kinked-out into the matrix at different debond lengths d_l. Source: Zhuang, L., Talreja, R. and Varna, J., 2016. Tensile failure of unidirectional composites from a local fracture plane. *Composites Science and Technology*, *133*, 119–127.

stress on fibers overcomes their strength. This statistical process of fiber breakage is facilitated by the growing matrix crack, which also provides a continuous, albeit erratic plane from which final failure results.

5.2.1 EFFECT OF MANUFACTURING DEFECTS

As discussed above, the first failure event in the failure process in UD composites under axial tension is a single fiber break due to its tensile stress exceeding the local fiber strength. That local fiber strength depends on the flaws in the fiber caused by its own manufacturing process, e.g., by causing fiber diameter variation, and those resulting from fiber surface treatment such as sizing that induces roughness on the fiber surface [12]. Surface treatments are necessary to protect fibers from damage during handling, e.g., in preform placement in RTM and in pre-tensioning in prepreg manufacture. The increase in statistical scatter in the tensile strength of fibers results in the early occurrence of singlets as the first failure event. At the same time, the failure process can also be started by pre-existing broken fibers. One study [13] compared the failure scenarios resulting from a pre-existing broken fiber versus when the breakage of the weakest fiber starts the process. Based on local stress analysis in the vicinity of a broken fiber end, the study found that a matrix crack emanating from a pre-existing broken fiber end tends to grow normal to the fiber axis and on reaching a neighboring fiber induces more stress concentration than that caused by the broken fiber. Since the fiber/matrix debonding is small compared to when the weakest fiber breaks by stressing, the matrix crack emanating from the pre-existing broken fiber end does not spread the stress enhancement on the neighboring fiber but localizes it where it hits that fiber. Such findings of a numerical study are difficult to verify by experimental observations because of the very small scale of the cracks involved.

The effect of matrix voids on fiber break initiation has not been studied much because of the focus on the fiber breakage process itself. One XCT study reported [14] that voids affected fiber breakage only within a short distance, approximately half the fiber diameter. Some of the voids were found to intersect with cracks in the matrix, suggesting that voids had an influence on the formation of cracks, but whether those cracks in turn influenced fiber breakage could not be determined.

5.3 FAILURE IN LONGITUDINAL COMPRESSION

In the absence of clear observations to indicate how failure in UD composites occurred under axial compression, the first modeling effort by Rosen [15] postulated that straight continuous fibers embedded in a matrix deflected as locally buckled columns. The buckling mode was assumed to be a waveform at a micrometer level, named microbuckling, and was assumed to occur either as all fibers buckling in-phase or out-of-phase modes. The in-phase microbuckling model produced a lower critical load, and was assumed to be more likely, but was still found to overestimate the experimentally found failure load. Later, Argon [16] explained that the Rosen model was possible only if the fibers were straight and parallel. In a real manufacturing process, fibers are likely to be misaligned, in which case, in the region of fiber misalignment a failure nucleus will form triggering a kinking process from the in-phase microbuckling mode. Thus, a connection with fiber misalignment, a manufacturing-induced defect, was theoretically established for the first time. In numerous experimental studies in the ensuing years, evidence emerged of the existence of kink bands associated with the failure of UD composites under axial compression. Prompted by that evidence, research in kink band formation and growth, and eventual failure, has been an active field, and numerous key papers and reviews of the field have appeared [17–28]. While modeling aspects will be discussed in a later chapter, we shall here highlight the nature of the compressive failure process and emphasize the roles of manufacturing-induced defects.

The pre-existing fiber misalignment can give rise to a nucleus for the formation of a kink band, as was suggested by Argon [16]. The experimental study conducted in [24] explored the nature of that nucleus and found that a pre-existing misaligned fiber was not the only source of the nucleating zone for kink band formation. The authors found evidence of initially broken or crushed fibers during processing to also be a source for kink band initiation, which resulted from local microbuckling of fibers. Once initiated, a kink band appears as depicted by an SEM image in Figure 5.5 [24]. In fact, the kink bands observed vary significantly depending on the constituent and interface properties in addition to the source triggering the initiation of the kink band. A general geometry of a kink band, often assumed in modeling studies, is depicted in Figure 5.6.

The kink band width W, its inclination angle β and the rotation of fibers within a kink band given by the sum of the initial fiber misalignment ϕ and the rotation ϕ_s

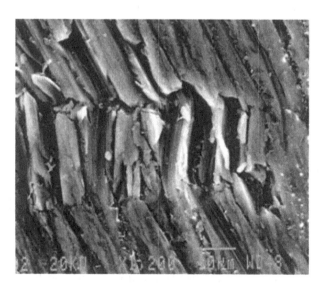

FIGURE 5.5   An SEM image of a kink band formed in a single tow carbon/epoxy composite. Source: Narayanan, S. and Schadler, L.S., 1999. Mechanisms of kink-band formation in graphite/epoxy composites: a micromechanical experimental study. *Composites Science and Technology*, *59*(15), 2201–2213.

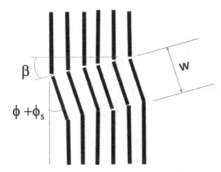

FIGURE 5.6 An idealized kink band geometry. W is the kink band width, b denotes the angle of kink band rotation, ϕ and ϕ_s are the initial fiber misalignment angle and the fiber rotation angle, respectively.

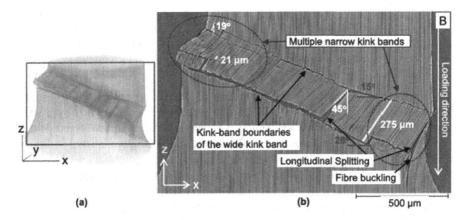

FIGURE 5.7 The zone of a partially failed specimen scanned by XCT (a), and (b) tomogram showing multiple kink bands, buckled and broken fibers as well as splitting along fibers in a kink band. Source: Wang, Y., Burnett, T.L., Chai, Y., Soutis, C., Hogg, P.J. and Withers, P.J., 2017. X-ray computed tomography study of kink bands in unidirectional composites. *Composite Structures*, *160*, 917–924.

caused by local shear have all been subjects of modeling and simulation. The width W of the kink band has been found to not always be constant, but increases as the band propagates [26]. More recently, high-resolution X-ray tomography has provided many details not included in the idealized geometry shown in Figure 5.6 [28]. The tomogram in Figure 5.7 displays the presence of multiple kink bands of different orientations in a damage zone of a partially failed specimen. It also shows with clarity buckled and broken fibers as well as splitting (matrix cracks) along fibers within a kink band.

When thick composites are manufactured, a common type of defect is the waviness of layers. To investigate how such defects influence compressive failure, the study [29] implanted waviness in a thick UD composite and systematically examined

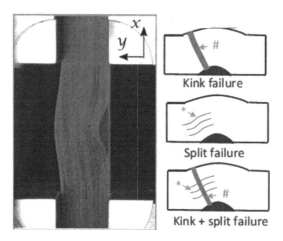

FIGURE 5.8   A thick carbon-epoxy UD composite with implanted waviness (left) and failure modes (right) depending on the waviness parameters. Source: Davidson, P. and Waas, A.M., 2017. The effects of defects on the compressive response of thick carbon composites: An experimental and computational study. *Composite Structures*, *176*, 582–596.

the failure behavior under an axial compression load. The specimen with implanted waviness is shown in Figure 5.8. The waviness geometry was defined by a sinusoid whose maximum undulation (slope), wavelength, and amplitude were taken as parameters. The failure modes observed were as schematically depicted in Figure 5.8, i.e., kinking, splitting (cracking between layers), and a combination of the two. The kinking mode correlated with the maximum undulation while splitting seemed to be influenced by the wave amplitude to length ratio.

Other than misaligned and broken fibers resulting from manufacturing processes, voids in the matrix have also been found to influence kink bands [30]. In this study, carbon/epoxy specimens with an inclined notch subjected to axial compression were observed under in situ SEM. The specimens were cut from a panel that was manufactured in an autoclave with reduced cure pressure to allow voids to remain in the matrix. The notch tip area was observed to follow failure events under compression. It was found that for kink bands forming near a void tip, the reduced lateral support to fibers caused by a void changed the inclination of the kink band. Also, a void in the path of a growing kink band was found to arrest the kink band.

5.4   FAILURE IN TRANSVERSE TENSION

Experimental observation of how failure occurs when a tensile load is applied transversely, i.e., normal to the fiber axis of a UD composite, is made difficult by the fact that the failure occurs abruptly. It is therefore common to observe the failure process by focusing on the 90° layers of a laminate loaded under tension in the 0-direction. In relatively thick layers of the 90° plies the region of the plies away from the interfaces with other plies is then in approximately uniform transverse tension

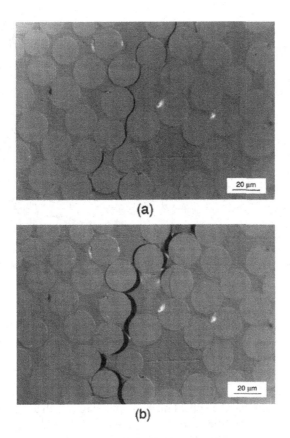

FIGURE 5.9 (a) Fiber/matrix debonding, and (b) coalescence of debonds into a transverse crack. Source: Gamstedt, E.K. and Sjögren, B.A., 1999. Micromechanisms in tension-compression fatigue of composite laminates containing transverse plies. *Composites Science and Technology, 59*(2), 167–178.

and the failure process is slowed down by the constraint of the other plies. A good example of this was reported in [31] from which Figure 5.9 is taken. The authors of this study manufactured a [0/90]$_s$ laminate with glass fibers and vinyl-ester resin by an RTM process with an outer 0° layer thickness of 0.5 mm and inner 90° layers of 1.0 mm. They applied a tensile load in the 0° direction and observed the edge of the specimen under an optical microscope. They found that the fibers debonded from the matrix initially, and as the load was increased, some of the debond cracks coalesced to form a transverse crack, as depicted in Figure 5.9(a). On further loading, a coalesced crack showed crack surface separation, indicating a continuous crack, as seen in Figure 5.9(b). Similar observations were reported earlier in [32] for a carbon-epoxy laminate. Later, the fiber/matrix debonding was reported at a higher resolution in [33] where an SEM image of a transverse crack and a zoomed-in image of a few fibers in a 90° layer of a 0/45/90 carbon-epoxy laminate were shown, Figure 5.10(a) and (b).

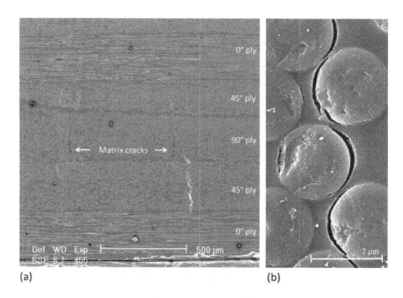

(a)   (b)

FIGURE 5.10 SEM images (a) of matrix cracks in a 90-degree ply of a 0/45/90 carbon-epoxy laminate, and (b) a zoomed-in image of a matrix crack showing fiber/matrix debonding. Source: Romanov, V.S., Lomov, S.V., Verpoest, I. and Gorbatikh, L., 2015. Modelling evidence of stress concentration mitigation at the micro-scale in polymer composites by the addition of carbon nanotubes. *Carbon*, *82*, 184–194.

It should be noted that the optical microscopy images such as those in Figure 5.8 and the higher-resolution SEM images in Figure 5.10 show the random location of the matrix cracks in the initial stages reflecting the non-uniformity of fiber distribution in the cross section caused by manufacturing. However, such images are not able to show the beginning of the fiber/matrix debonding process, which occurs at a much lower scale than what such images can capture. It is common to assume that the debonding occurs by breaking the interface bonds. One study [34] investigated the possibility that the matrix itself could fail near the fiber/matrix interface before any debonding. The explanation given was that under transverse tension the stress state near the fiber surfaces at some points becomes nearly hydrostatic tension leading to brittle cavitation, i.e., failure by volumetric expansion without inelasticity. Under equi-triaxial tension, then, the strain to failure is much smaller than what it would be under uniaxial stress. One example to illustrate this phenomenon is in Figure 5.11, which shows the stress–strain plots of an epoxy under uniaxial tension and under triaxial tension [34]. The triaxial test was conducted by what is known as a poker-chip test [34]. The repercussions of the brittle cavitation on modeling of transverse failure in composites will be described in detail in a later chapter when failure analysis is discussed.

Manufacturing-induced voids were discussed in Chapter 4. These voids can be initiators of fiber/matrix debonding under transverse tension in UD composites. One clear evidence of this mechanism is indicated in Figure 5.12 [35]. Figure 5.13(a)

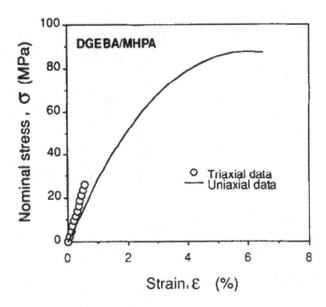

FIGURE 5.11 Experimental stress-strain data for an epoxy under uniaxial and triaxial (hydrostatic) tension. Source: Asp, L.E., Berglund, L.A. and Gudmundson, P., 1995. Effects of a composite-like stress state on the fracture of epoxies. *Composites Science and Technology*, *53*(1), 27–37.

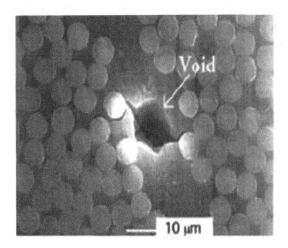

FIGURE 5.12 A matrix void initiating fiber/matrix debonding under transverse tension. Source: Wood, C.A. and Bradley, W.L., 1997. Determination of the effect of seawater on the interfacial strength of an interlayer E-glass/graphite/epoxy composite by in situ observation of transverse cracking in an environmental SEM. *Composites Science and Technology*, *57*(8), 1033–1043.

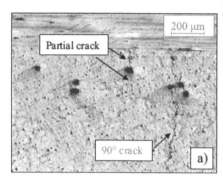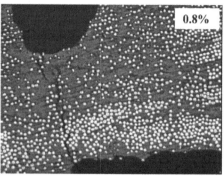

FIGURE 5.13 Left: Transverse cracks emanating from voids in 90° layer of a laminate under axial (horizontal) tension, and right: Details of a crack connecting two neighboring voids at an applied transverse strain of 0.8%. Source: (left figure): Maragoni, L., Carraro, P.A., Peron, M. and Quaresimin, M., 2017. Fatigue behaviour of glass/epoxy laminates in the presence of voids. *International journal of fatigue*, 95, 18–28., (right figure): Huang, Y., Varna, J. and Talreja, R., 2014. Statistical methodology for assessing manufacturing quality related to transverse cracking in cross ply laminates. *Composites Science and Technology*, 95, 100–106.

[36] shows transverse cracks that are connected to matrix voids in the 90° plies of a laminate. These cracks were found to emanate from voids and grow by connecting neighboring voids, if available. One case of a transverse crack connecting two voids is illustrated in Figure 5.13 (b) [37]. Such effects of voids on matrix crack initiation work in combination with the presence of non-uniform fiber distribution in the UD composite cross section.

5.5 FAILURE IN TRANSVERSE COMPRESSION

UD composites typically have much higher strength in transverse compression than in transverse tension, and failure in transverse compression seldom is the limiting design consideration. For these reasons, this failure mode has not been studied extensively. One study [38] looked at the failure process by taking SEM images of the failure plane and these have been reported in [39], Figure 5.14 shows those images. As seen, the failure appears to have come from a plane (shear band) inclined to the loading axis, suggesting that it has been largely shear driven. The close-up image taken near the failure plane shows fiber/matrix debonding along the shear band. Finally, the SEM image of the fracture plane shows hackles confirming the shearing of the matrix.

Post-failure observations such as those in Figure 5.14 do not provide direct evidence of the first failure event and the early progression of failure. A study [40] attempted to achieve this by testing a carbon-epoxy UD composite under compressive loads of 25%, 50%, 75%, and 80% of the average failure load and observing the polished specimen edge under an optical microscope. Based on those observations the authors described the failure process as occurring in stages of fiber/matrix

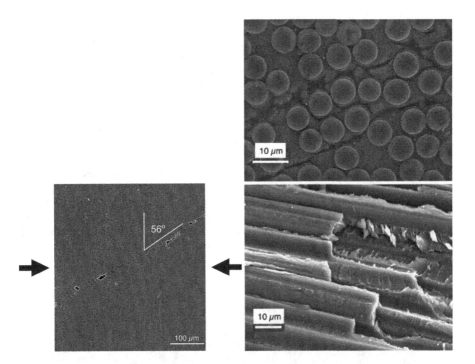

FIGURE 5.14 Left: SEM image of the lateral surface of a carbon-epoxy UD composite taken before the failure load showing the formation of a failure plane inclined to the loading direction indicated by arrows. Top: a close-up of the lateral surface near the failure plane showing interface debonding. Right: Fracture surface showing the presence of matrix shear (hackles). Source: González, C. and LLorca, J., 2007. Mechanical behavior of unidirectional fiber-reinforced polymers under transverse compression: Microscopic mechanisms and modeling. *Composites Science and Technology*, *67*(13), 2795–2806.

interface debonding and debond crack kinking culminating in a macro-crack that leads to final failure. Their observations are summarized in Figure 5.15. It is noted that these images are selected from different specimens to illustrate the failure advanced to increasing levels prior to final failure.

A recent study [41] took pains to machine out small samples from UD composites of carbon/epoxy where direct SEM observations could be made at the fiber scale. This study found evidence of fiber/matrix debonding followed by matrix cracks in the loading direction that connected the debonds. The final failure occurred on planes inclined to the loading direction (Figure 5.16). These observations will be discussed in Chapter 8 with respect to the failure modeling efforts.

There are no experimental observations of the transverse compression failure in UD composites that clarify how manufacturing defects (non-uniform fiber distribution in the cross section and matrix voids) affect the failure process. The modeling and simulation studies are based on assumptions regarding the initiation and progression of failure, and those will be discussed in a later chapter.

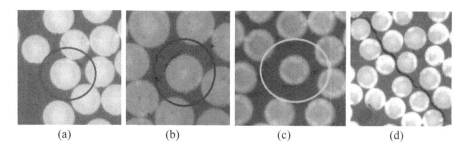

(a)   (b)   (c)   (d)

FIGURE 5.15 Optical micrographs of a UD carbon-epoxy composite under transverse compression in the vertical direction were taken to illustrate evidence of different stages of failure. (a) Initiation of fiber/matrix debond, (b) growth of debond, (c) kink-out of a debond crack, and (d) macro-crack before final failure. Source: Zumaquero, P.L., Correa, E., Justo, J. and París, F., 2018. Microscopical observations of interface cracks from inter-fibre failure under compression in composite laminates. *Composites Part A: Applied Science and Manufacturing, 110*, 76–83.

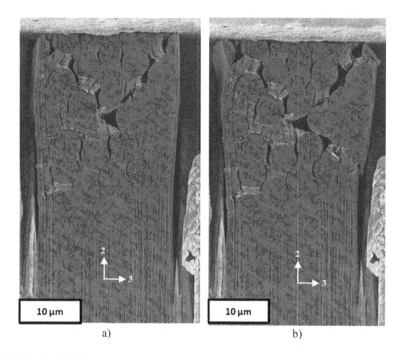

a)   b)

FIGURE 5.16 SEM images of a microscopic specimen loaded in compression in the transverse (vertical) direction. (a) Prior to ultimate failure and (b) after ultimate failure. Source: Flores, M., Sharits, A., Wheeler, R., Sesar, N. and Mollenhauer, D., 2022. Experimental analysis of polymer matrix composite microstructures under transverse compression loading. *Composites Part A: Applied Science and Manufacturing, 156*, 106859.

5.6 FAILURE IN IN-PLANE SHEAR

Inducing a uniform state of shear stress in a UD composite faces practical difficulties. Various methods have been proposed, and a comparison of the common methods, e.g., panel-shear, rail-shear, +45/−45 axial tension, 10° off-axis tension, and tube under torque produce different strength values [42, 43]. To avoid failure from grips, the Iosipescu test was developed, but it does not generate a pure shear stress state [44]. For producing pure shear, the preferred specimen is a thin-walled tube in torsion with the fibers running in the circumferential (hoop) direction. This testing method has been used by [45, 46] to study failure mechanisms in UD composites under in-plane shear. Their microscopic observations of failure induced by cyclic inplane shear confirmed earlier observations reported in [47, 48]. In [47] shear stress was introduced directly at the fiber level by cutting fibers that retracted inducing equal and opposite forces locally and in [48] shear failure was induced between fibers in a 10° off-axis tension fatigue. As seen in Figure 5.17, the shear failure evident from hackles on the fracture surface results in micro-cracks in the matrix along

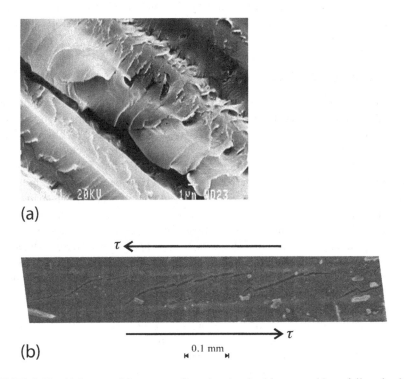

FIGURE 5.17   (a) Image of fracture surface showing hackles caused by a failure in shear, and (b) sigmoidal shape cracks in the matrix produced by axial shear between fibers. Source: (a): Plumtree, A. and Shi, L., 2002. Fatigue damage evolution in off-axis unidirectional CFRP. *International Journal of Fatigue*, 24(2–4), 155–159, (b) Redon, O., Fatigue damage development and failure in unidirectional and angle-ply glass fibre/carbon fibre hybrid laminates, Technical Report Risø-R-1168, 2000, Risø National Laboratory, Roskilde, Denmark.

planes inclined to the fibers. These cracks tend to turn and grow in the fiber direction, merging to form sigmoidal shape "axial" cracks.

5.7 FAILURE IN COMBINED LOADING

It is commonly assumed that when two elementary loading modes are applied together, e.g., longitudinal tension and in-plane shear, the failure mechanisms induced by the two modes individually undergo interaction. The nature of the interaction is, however, not always clear. In each loading mode, the failure process consists of a sequence of failure events, as described above, and the question to ask is: How are the failure events and their progression corresponding to one loading mode affected by those corresponding to the other loading mode? In the following, we shall address this question, taking two loading modes at a time. In doing so, we shall highlight the effects induced by the presence of manufacturing defects. It is noted that most studies of composite failure interaction have been focused on the final event of failure (strength) motivated by the need to describe a failure surface in the stress space in the context of a failure theory. The multiaxial testing for this purpose is fraught with practical issues of specimen shapes and gripping fixtures. A survey of the test methods can be found in [49].

5.7.1 LONGITUDINAL TENSION AND IN-PLANE SHEAR

As discussed above in Section 5.2, the failure events in UD composites under longitudinal tension occur at the fiber scale, which necessitates using high-resolution techniques such as XCT to get sufficient clarity of their nature. Yet, observations of the fiber breakage event even with this technique do not provide clarification of the consequences of those events on the local deformation and/or failure of the matrix and the fiber/matrix interface. These consequences have been examined by modeling and simulation studies where the results depend on the assumptions made on the matrix and interface properties. It is expected that applying shear in the UD composite plane will affect the matrix and interface consequences of fiber failures caused by the longitudinal tension if the shear itself does not induce failure. It turns out, however, that at low values of the applied shear stress, axial splitting tends to occur making it difficult to study the effect of shear on failure induced by longitudinal tension. On the other hand, observing local failure events caused by in-plane shear is not easy, as discussed above, and applying longitudinal tension can make matters worse if fibers break by longitudinal tension. As a result, not much has been reported in the literature on the combined effects of longitudinal tension and in-plane shear on the failure of UD composites. Discerning any interaction between failure caused by the two loading modes individually from the final failure results is also uncertain because of the significant scatter in the strength data that is inherent in the tensile failure.

Due to the difficulties of following the failure events under combined longitudinal tension and in-plane shear, the effect of manufacturing defects on the failure progression has not been reported. It is conceivable that the presence of matrix voids will enhance local shear in the vicinity of voids, thereby affecting the local load sharing between fibers. A computational study [50] has suggested that the shear yielding of the matrix reduces the longitudinal stress in fibers. This would imply that matrix voids will have a favorable effect on the composite failure from fiber breakage while increasing the propensity of matrix failure.

5.7.2 LONGITUDINAL TENSION AND TRANSVERSE TENSION

The difference in the strength of UD composites in the fiber direction and in the transverse direction is usually large for glass-epoxy and carbon-epoxy composites. This limits the range of combined loading in which interaction between failures in the two loading modes can be observed. The final failure (strength) data [51] suggest failure either by fiber failure in longitudinal tension or by unstable crack growth along fibers in transverse tension. The apparent lack of interaction between the two failure processes does not motivate the examination of the effects of manufacturing defects under combined loading.

5.7.3 LONGITUDINAL TENSION AND TRANSVERSE COMPRESSION

The strength values of UD composites in longitudinal tension and transverse compression are large [51] providing a good range of combined loading in which the interaction between failures in the two loading modes can be studied. However, test fixtures for applying the combined loading limit observations and much information concerning failure mode interaction are not available. It is conceivable that under longitudinal tension the local load sharing between broken fibers and their intact neighbors is affected by transverse compression via plastic deformation in the matrix at sufficiently high compressive stresses. The interactive effect then would be like the case of applying in-plane shear in addition to the longitudinal tension, although to a lesser degree. In the absence of observations at the microscopic level it has become common to study the combined loading interactions by virtual testing. It is then imperative that experimental verification is pursued to gain confidence in the predictions. The same goes for studies to investigate the effects of manufacturing defects on the loading mode interactions on failure.

5.7.4 LONGITUDINAL COMPRESSION AND IN-PLANE SHEAR

Failure under pure longitudinal compression has been found to induce kink bands, as discussed above. This failure is typically catastrophic and therefore it is difficult to track the growth of kink bands. In a study of combined longitudinal compression and in-plane shear [26] using a fixture developed for the purpose, the authors found that the kink band growth could be stabilized to follow its growth. Figure 5.16 depicts the geometry of the kink band as it grows across the fibers. Contrary to the case of pure compression, the kink band in combined loading was found to have few broken fibers. As depicted in Figure 5.18, the fiber curvature and rotation at the front

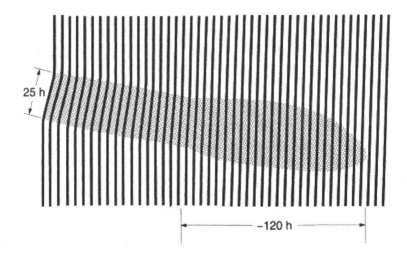

FIGURE 5.18   A sketch of a kink band growing in combined longitudinal compression and in-plane shear. The dimensions shown are in terms of the fiber diameter h. Source: Vogler, T.J. and Kyriakidis, S., 2001. On the initiation and growth of kink bands in fiber composites: Part 1. Experiments. *International Journal of Solids and Structures*, 38, 2639–2651.

edge of the kink band reduce to zero, making the applied compression insufficient to buckle the fibers. The kink band then grows due to the transverse displacement induced by shear. The study [52] performed XCT to examine the effect of combining shear with longitudinal compression for carbon-epoxy UD composites made of non-crimp fabric by testing specimens cut at off-axis angles and tested under longitudinal compression. It was found that kink bands formed out-of-plane when no shear was present but became increasingly oriented in the plane as the shear (off-axis angle) was increased.

Manufacturing defects such as misaligned and wavy fibers as well as matrix voids play roles in initiating kink bands, as described above for the case of pure longitudinal compression. Combining compression with in-plane shear can make the conditions for kink band initiation more favorable by adding to fiber rotation and matrix shear, and by increasing the shear stress concentration around voids.

5.7.5   LONGITUDINAL COMPRESSION AND TRANSVERSE TENSION

This case has also not been studied experimentally because of the low stress to failure in transverse tension compared to the longitudinal compressive strength. It can be expected that any fiber/matrix debonding under transverse tension is likely to reduce the lateral support to the fibers and thereby cause earlier fiber microbuckling. The splitting failures occurring near wavy layers under longitudinal compression reported in [29] are likely to grow when transverse tension is applied. An additional deleterious effect of transverse tension on longitudinal compression failure would be increasing stress concentration at matrix voids.

5.7.6 LONGITUDINAL COMPRESSION AND TRANSVERSE COMPRESSION

The effect of multiaxial compression was investigated in one study [53] where an E-glass/vinyl ester UD composite was examined after failure. The observations showed that a failure mode transition occurred from axial splitting to kink band formation going from uniaxial to multiaxial compression. Under longitudinal compression axial splitting-induced kink bands were found while under proportional multiaxial compression, only kink bands were observed.

The type of post-mortem observations reported in [53] are useful in the absence of in-situ tracking of failure events. The limitation of in-situ observations also leaves open questions concerning the effect of fiber misalignment and matrix voids on the failure process in combined compression.

5.7.7 TRANSVERSE TENSION AND IN-PLANE SHEAR

The specimen geometry most suited for this combined loading is a thin-walled cylinder with circumferential fibers loaded in axial tension and torsion. This specimen geometry was used in [45] to study failure initiation and progression under cyclic loading. By choosing the level of loading in the two modes and their relative values, the failure process could be studied in detail. To study the failure process in its entirety it was found useful to constrain the process by adding longitudinal fabrics to the tubes. Observations of the failure process were made by an infrared camera and visual inspection aided by lighting in the interior of the tubular specimen. The first failure event was assessed to be crack initiation along fibers, and it was found to be affected by the presence of defects. Once a crack was initiated its growth was in mixed mode as depicted in Figure 5.19 [45].

The effect of manufacturing defects such as non-uniform fiber distribution and matrix voids on transverse crack initiation was discussed above in Section 5.4. The local stress states in inter-fiber regions due to these defects will have stress concentrations and gradients under combined loading different from when only transverse tension is applied. Nevertheless, the crack initiation will be caused by these stress states. Evidence of crack initiation from voids in 45° plies of [+45/–45/0]$_s$ laminates under axial tension was reported in [54].

5.7.8 TRANSVERSE COMPRESSION AND IN-PLANE SHEAR

As described above, the failure in the two loading modes applied individually has been found to be dominated by shear-driven inelasticity. When the two loading modes are combined, the local stress states will depend on the load combination and consequently, the inelastic deformation in the inter-fiber regions will govern failure. An experimental study [55] tested carbon-PEEK composites in a bi-axial shear/compression device and found differences in the mechanical response in different loading paths. This can be attributed to the non-linear inelastic response. The failure mechanisms as observed on fracture surfaces in pure compression, pure shear, and combined compression/shear were found to be similar. The hackles found on the fracture surfaces were more for combined loading than for pure loading cases.

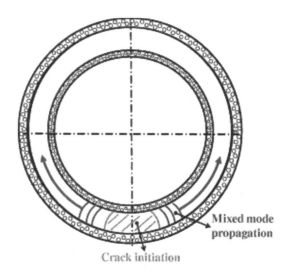

FIGURE 5.19   Crack initiation and its mixed-mode propagation along circumferential fibers in a tubular specimen subjected to longitudinal tension and torsion. Source: Carraro, P.A. and Quaresimin, M., 2014. A damage based model for crack initiation in unidirectional composites under multiaxial cyclic loading. *Composites Science and Technology*, *99*, 154–163.

5.8   CONCLUSION

This chapter has reviewed failure mechanisms in unidirectional composites under different loading modes. The initiation of failure, the progression of the failure events, and the critical states causing final failure have been described. The review has been organized to follow the reported testing and observations, i.e., in individual loading modes of tension and compression in longitudinal and transverse directions, in in-plane shear, and in combinations of these. In some of the cases, the failure events have not been clarified by in-situ observations due to difficulties in making such observations. This subjects the modeling efforts to some constraints and the assumptions made are then to be verified against what has been observed and measured experimentally. The challenges in this respect are significant when the effect of manufacturing defects must be assessed. Further discussion of this subject will be conducted in a later chapter on modeling.

REFERENCES

1. Fuwa, M., Bunsell, A.R. and Harris, B., 1975. Tensile failure mechanisms in carbon fibre reinforced plastics. *Journal of Materials Science*, *10*, pp. 2062–2070.
2. Purslow, D., 1981. Some fundamental aspects of composites fractography. *Composites*, *12*(4), pp. 241–247.
3. Aroush, D.R.B., Maire, E., Gauthier, C., Youssef, S., Cloetens, P. and Wagner, H.D., 2006. A study of fracture of unidirectional composites using in situ high-resolution synchrotron X-ray microtomography. *Composites Science and Technology*, *66*(10), pp. 1348–1353.

4. Scott, A.E., Mavrogordato, M., Wright, P., Sinclair, I. and Spearing, S.M., 2011. In situ fibre fracture measurement in carbon–epoxy laminates using high resolution computed tomography. *Composites Science and Technology, 71*(12), pp. 1471–1477.
5. Scott, A.E., Sinclair, I., Spearing, S.M., Thionnet, A. and Bunsell, A.R., 2012. Damage accumulation in a carbon/epoxy composite: Comparison between a multiscale model and computed tomography experimental results. *Composites Part A: Applied Science and Manufacturing, 43*(9), pp. 1514–1522.
6. Garcea, S.C., Sinclair, I., Spearing, S.M. and Withers, P.J., 2017. Mapping fibre failure in situ in carbon fibre reinforced polymers by fast synchrotron X-ray computed tomography. *Composites Science and Technology, 149*, pp. 81–89.
7. Swolfs, Y., Morton, H., Scott, A.E., Gorbatikh, L., Reed, P.A., Sinclair, I., Spearing, S.M. and Verpoest, I., 2015. Synchrotron radiation computed tomography for experimental validation of a tensile strength model for unidirectional fibre-reinforced composites. *Composites Part A: Applied Science and Manufacturing, 77*, pp. 106–113.
8. Garcia, S.C., Wang, Y. and Withers, P.J., 2018. X-ray computed tomography of polymer composites. *Composites Science and Technology, 156*, pp. 305–319.
9. Na, W., Kwon, D. and Yu, W.R., 2018. X-ray computed tomography observation of multiple fiber fracture in unidirectional CFRP under tensile loading. *Composite Structures, 188*, pp. 39–47.
10. Schöberl, E., Breite, C., Melnikov, A., Swolfs, Y., Mavrogordato, M., Sinclair, I. and Spearing, S.M., 2020. Fibre-direction strain measurement in a composite ply under quasi-static tensile loading using Digital Volume Correlation and *in situ* synchrotron radiation computed tomography. *Composites Part A: Applied Science and Manufacturing, 137*, 105935.
11. Phoenix, S.L. and Beyerlein, I.J., 2000. Statistical strength theory for fiber composite materials. In Kelly, A. (ed.), *Comprehensive Composite Materials*. Elsevier, Amsterdam, Chapter 19, pp. 559–639.
12. Zhao, F.M. and Takeda, N., 2000. Effect of interfacial adhesion and statistical fiber strength on tensile strength of unidirectional glass fiber/epoxy composites. Part I: experiment results. *Composites Part A: Applied Science and Manufacturing, 31*(11), pp. 1203–1214.
13. Zhuang, L., Talreja, R. and Varna, J., 2016. Tensile failure of unidirectional composites from a local fracture plane. *Composites Science and Technology, 133*, pp. 119–127.
14. Scott, A.E., Sinclair, I., Spearing, S.M., Mavrogordato, M.N. and Hepples, W., 2014. Influence of voids on damage mechanisms in carbon/epoxy composites determined via high resolution computed tomography. *Composites Science and Technology, 90*, pp. 147–153.
15. Rosen, V.W., 1965. Mechanics of composite strengthening. In *Fiber Composite Materials*. American Society of Metals, Metals Park, Ohio, pp. 37–75.
16. Argon, A.S., 1972. Fracture of composites. In *Treatise of Materials Science and Technology*, Vol. 1. Academic Pres, New York, pp. 79–114.
17. Schultheisz, C.R. and Waas, A.M., 1996. Compressive failure of composites, part I: testing and micromechanical theories. *Progress in Aerospace Sciences, 32*(1), pp. 1–42.
18. Budiansky, B., 1983. Micromechanics. *Computers & Structures, 16*(1–4), pp. 3–12.
19. Componeshi, E., 1991. Compression of composite materials- A review. *Composite materials: Fatigue and Fracture, 3*, pp. 550–578.
20. Fleck, N.A., 1997. Compressive failure of fiber composites. In Hutchinson, J.W. and Wu, T.Y. (eds.), *Advances in Applied Mechanics*, Vol. 33. Academic Press, New York, pp. 47–117.
21. Soutis, C.,1997. Compressive strength of unidirectional composites: Measurement and prediction. In Hooper, S.J. (ed.), *Composite Materials: Testing and Design*, Vol. 13. ASTM STP 1242, Philadelphia, PA, pp. 168–176.

22. Vogler, T.J. and Kyriakides, S., 1997. Initiation and axial propagation of kink bands in fiber composites. *Acta Materialia*, *45*(6), pp. 2443–2454.
23. Jensen, H.M. and Christoffersen, J., 1997. Kink band formation in fiber reinforced materials. *Journal of the Mechanics and Physics of Solids*, *45*(7), pp. 1121–1136.
24. Narayanan, S. and Schadler, L.S., 1999. Mechanisms of kink-band formation in graphite/epoxy composites: A micromechanical experimental study. *Composites Science and Technology*, *59*(15), pp. 2201–2213.
25. Niu, K. and Talreja, R., 2000. Modeling of compressive failure in fiber reinforced composites. *International Journal of Solids and Structures*, *37*(17), pp. 2405–2428.
26. Vogler, T.J. and Kyriakidis, S., 2001. On the initiation and growth of kink bands in fiber composites: Part 1. Experiments. *International Journal of Solids and Structures*, *38*, pp. 2639–2651.
27. Pimenta, S., Gutkin, R., Pinho, S.T. and Robinson, P., 2009. A micromechanical model for kink-band formation: Part I—Experimental study and numerical modelling. *Composites Science and Technology*, *69*(7–8), pp. 948–955.
28. Wang, Y., Burnett, T.L., Chai, Y., Soutis, C., Hogg, P.J. and Withers, P.J., 2017. X-ray computed tomography study of kink bands in unidirectional composites. *Composite Structures*, *160*, pp. 917–924.
29. Davidson, P. and Waas, A.M., 2017. The effects of defects on the compressive response of thick carbon composites: An experimental and computational study. *Composite Structures*, *176*, pp. 582–596.
30. Hapke, J., Gehrig, F., Huber, N., Schulte, K. and Lilleodden, E.T., 2011. Compressive failure of UD-CFRP containing void defects: In situ SEM microanalysis. *Composites Science and Technology*, *71*(9), pp. 1242–1249.
31. Gamstedt, E.K. and Sjögren, B.A., 1999. Micromechanisms in tension-compression fatigue of composite laminates containing transverse plies. *Composites Science and Technology*, *59*(2), pp. 167–178.
32. Harrison, R.P. and Bader, M.G., 1983. Damage development in CFRP laminates under monotonic and cyclic stressing. *Fibre Science and Technology*, *18*, pp. 163–180.
33. Romanov, V.S., Lomov, S.V., Verpoest, I. and Gorbatikh, L., 2015. Modelling evidence of stress concentration mitigation at the micro-scale in polymer composites by the addition of carbon nanotubes. *Carbon*, *82*, pp. 184–194.
34. Asp, L.E., Berglund, L.A. and Gudmundson, P., 1995. Effects of a composite-like stress state on the fracture of epoxies. *Composites Science and Technology*, *53*(1), pp. 27–37.
35. Wood, C.A. and Bradley, W.L., 1997. Determination of the effect of seawater on the interfacial strength of an interlayer E-glass/graphite/epoxy composite by in situ observation of transverse cracking in an environmental SEM. *Composites Science and Technology*, *57*(8), pp. 1033–1043.
36. Maragoni, L., Carraro, P.A., Peron, M. and Quaresimin, M., 2017. Fatigue behaviour of glass/epoxy laminates in the presence of voids. *International Journal of Fatigue*, *95*, pp. 18–28.
37. Huang, Y., Varna, J. and Talreja, R., 2014. Statistical methodology for assessing manufacturing quality related to transverse cracking in cross ply laminates. *Composites Science and Technology*, *95*, pp. 100–106.
38. Argones, D., 2007. Fracture micromechanisms in C/epoxy composites under transverse compression. Masters thesis, Polytechnical University of Madrid.
39. González, C. and LLorca, J., 2007. Mechanical behavior of unidirectional fiber-reinforced polymers under transverse compression: Microscopic mechanisms and modeling. *Composites Science and Technology*, *67*(13), pp. 2795–2806.
40. Zumaquero, P.L., Correa, E., Justo, J. and París, F., 2018. Microscopical observations of interface cracks from inter-fibre failure under compression in composite laminates. *Composites Part A: Applied Science and Manufacturing*, *110*, pp. 76–83.

41. Flores, M., Sharits, A., Wheeler, R., Sesar, N. and Mollenhauer, D., 2022. Experimental analysis of polymer matrix composite microstructures under transverse compression loading. *Composites Part A: Applied Science and Manufacturing*, 156, 106859.
42. Terry, G., 1979. A comparative investigation of some methods of unidirectional, in-plane shear characterization of composite materials. *Composites*, 10, pp. 233–237.
43. Van Paepegem, W., De Baere, I. and Degrieck, J., 2006. Modelling the nonlinear shear stress–strain response of glass fibre-reinforced composites. Part I: Experimental results. *Composites Science and Technology*, 66(10), pp. 1455–1464.
44. Pierron, F. and Vautrin, A., 1997. New ideas on the measurement of the in-plane shear strength of unidirectional composites. *Journal of Composite Materials*, 31(9), pp. 889–895.
45. Carraro, P.A. and Quaresimin, M., 2014. A damage based model for crack initiation in unidirectional composites under multiaxial cyclic loading. *Composites Science and Technology*, 99, pp. 154–163.
46. Quaresimin, M. and Carraro, P.A., 2014. Damage initiation and evolution in glass/epoxy tubes subjected to combined tension–torsion fatigue loading. *International Journal of Fatigue*, 63, pp. 25–35.
47. Redon, O., 2000. Fatigue damage development and failure in unidirectional and angle-ply glass fibre/carbon fibre hybrid laminates. Technical Report Risø-R-1168, Risø National Laboratory, Roskilde.
48. Plumtree, A. and Shi, L., 2002. Fatigue damage evolution in off-axis unidirectional CFRP. *International Journal of Fatigue*, 24(2–4), pp. 155–159.
49. Olsson, R., 2011. A survey of test methods for multiaxial and out-of-plane strength of composite laminates. *Composites Science and Technology*, 71(6), pp. 773–783.
50. Behzadi, S., Curtis, P.T. and Jones, F.R., 2009. Improving the prediction of tensile failure in unidirectional fibre composites by introducing matrix shear yielding. *Composites Science and Technology*, 69(14), pp. 2421–2427.
51. Hinton, M.J., Kaddour, A.S. and Soden, P.D., 2002. A comparison of the predictive capabilities of current failure theories for composite laminates, judged against experimental evidence. *Composites Science and Technology*, 62(12–13), pp. 1725–1797.
52. Wilhelmsson, D., Mikkelsen, L.P., Fæster, S. and Asp, L.E., 2019. Influence of in-plane shear on kink-plane orientation in a unidirectional fibre composite. *Composites Part A: Applied Science and Manufacturing*, 119, pp. 283–290.
53. Oguni, K., Tan, C.Y. and Ravichandran, G., 2000. Failure mode transition in unidirectional E-glass/vinylester composites under multiaxial compression. *Journal of Composite Materials*, 34(24), pp. 2081–2097.
54. Maragoni, L., Carraro, P.A. and Quaresimin, M., 2016. Effect of voids on the crack formation in a [45/−45/0] s laminate under cyclic axial tension. *Composites Part A: Applied Science and Manufacturing*, 91, pp. 493–500.
55. Vogler, T.J. and Kyriakides, S., 1999. Inelastic behavior of an AS4/PEEK composite under combined transverse compression and shear. Part I: experiments. *International Journal of Plasticity*, 15(8), pp. 783–806.

6 Failure in Laminates in the Presence of Defects

6.1 CONSTRAINED PLY FAILURE

If a single isolated unidirectional (UD) composite layer is loaded in axial tension or compression, or in in-plane shear, the failure process initiated by the loading is observed to be dependent on the loading mode, as described in the previous chapter. When the same layer is within a laminate, the externally applied loading modes will induce different stress fields in the constituents and at their interfaces. Therefore, the consequences of the loading modes in initiating the first failure event and subsequent evolution of failure are expected to be different in a free (isolated) UD composite versus in the same composite constrained within a laminate. Additionally, failure occurring at the ply/ply interfaces in a laminate is obviously not present in a free UD composite. This failure is commonly described as delamination because it leads to an apparent separation of two layers (laminae) at their interface. The delamination failure and the failure within the layers are observed to influence each other. To describe the experimental observations in a systematic manner, it is convenient to describe the intra-ply failure and inter-ply failure separately. This section will focus on the observations of failure within a ply that deforms under the constraints imposed by other plies of a laminate.

The damage mechanisms observed without the influence of manufacturing defects will be reviewed first, following which matrix and fiber-related defects will be considered.

6.2 LOADING MODE: AXIAL TENSION

6.2.1 Cross-Ply Laminates

The simplest laminate configuration for observing the failure process in a UD composite under the constraint of other UD composites bonded to it is the cross-ply laminate. Historically, the multiple ply cracking process was discovered as an essential outcome of the constraint effect by observing a glass fiber-reinforced polymer cross-ply laminate under axial tension [1]. One of the earliest observations depicting the constraint effect in terms of the relative thickness of the cracking 90° ply with respect to the thickness of the constraining 0° plies is shown in Figure 6.1 [2]. As can be seen in the figure, the degree of multiple cracking, measured in terms of the inverse of crack spacing (called crack number density, or simply, crack density) increases with the ply constraint. As a limiting condition of reducing the constraint, failure will occur from a single crack.

DOI: 10.1201/9781003225737-6

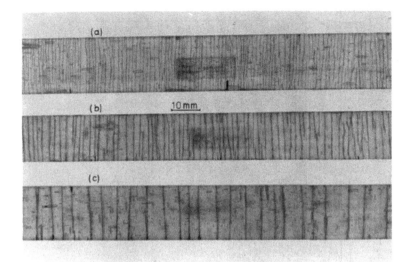

FIGURE 6.1 Multiple cracking in the transverse plies of cross-ply laminates of glass fiber–reinforced polymer composites under axial tension. Source: Garrett, K.W. and Bailey, J.E., 1977. Multiple transverse fracture in 90° cross-ply laminates of a glass fibre-reinforced polyester. *Journal of Materials Science*, 12, 157–168.

The constraint effect to multiple cracking can also be expressed in terms of the applied axial strain at which the cracking process begins. This approach was used in [3] to describe the level of constraint, labeled as A, B, C, and D with strain limits of fiber failure and single transverse crack formation, as indicated in Figure 6.2.

The initiation of crack formation has been of interest. One study [4] observed the early stage of the ply cracking process as influenced by the 0/90 ply interfaces. The authors tested $[0/90_n]_s$ laminates of carbon/epoxy under axial tension and observed the free edges of the laminates to capture the early initiation of the transverse cracks. Figure 6.3 from that study shows evidence of cracks initiating from the 0/90 ply interfaces before forming cracks that are then described by the multiple cracking measures such as the crack density. It is not clear whether the initiation of cracks from the interfaces is due to the local stress state at the free edges or if the local defects such as resin-rich areas and non-uniform fiber distribution in the interfaces are responsible for the initiation of cracks.

The effect of manufacturing defects on the multiple cracking evolution in cross-ply laminates was studied in [5]. In that study, panels of $[0/90]_s$ laminates were produced in an autoclave under four curing conditions: 1) without applying vacuum and pressure (NV_NP), 2) without vacuum but with 3 bars of pressure (NV_P), 3) with vacuum but no pressure (V_NP), and 4) with supplier specified vacuum and pressure (V_P). Cracks were observed on the free edges and counted over a 70 mm gauge length. The crack density (cracks/mm) evolution in the four cases is shown plotted in Figure 6.4. As seen in the figure, the curing with no vacuum and no pressure (NV_NP) leads to most cracks compared to the standard curing case, while

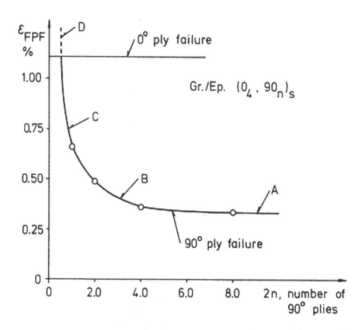

FIGURE 6.2 Strain to first ply failure ε_{FPF} in a cross-ply laminate varying with the constraining effect. Source: Talreja, R., 1985. Transverse cracking and stiffness reduction in composite laminates. *Journal of Composite Materials* 19, 355–375.

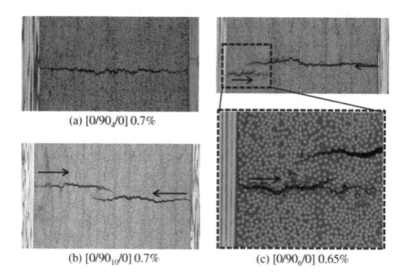

FIGURE 6.3 Initial stages of crack formation in 90° plies of different cross-ply laminates subjected to axial tension. The strain levels measured by DIC are indicated. Source: Okabe, T., Imamura, H., Sato, Y., Higuchi, R., Koyanagi, J. and Talreja, R., 2015. Experimental and numerical studies of initial cracking in CFRP cross-ply laminates. *Composites Part A: Applied Science and Manufacturing*, 68, 81–89.

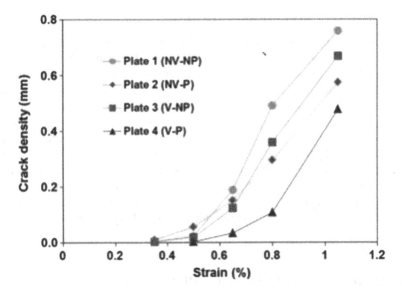

FIGURE 6.4 Crack density (cracks/mm) plotted against the applied axial strain in $[0/90]_s$ carbon/epoxy laminated manufactured in four curing conditions. Source: Huang, Y., Varna, J. and Talreja, R., 2014. Statistical methodology for assessing manufacturing quality related to transverse cracking in cross ply laminates. *Composites science and technology*, 95, 100–106.

the crack density in the other two non-standard conditions falls in between the two. Microscopy observations showed that voids induced by non-standard curing were initiators of cracks in the 90° plies.

6.2.2 GENERAL LAMINATES

While cross-ply laminates are suited for clarity of the constraint effects on intralaminar and interlaminar cracking, these laminates are not common in applications. Other configurations of laminates containing off-axis angles of multiple orientations are more common. A systematic study of constrained ply cracking varying the off-axis angle while keeping the axial plies as constraining plies was performed in [6]. In that study, laminates of $[0/+\theta/-\theta/0_{1/2}]_s$ layup tested in axial tension showed that as the angle θ was decreased from 90°, the initiation of ply cracks was delayed, and the crack density reached at the same applied load was also lowered. In fact, for θ approaching 0°, ply cracks were not seen before the failure of the laminate. Axial splits and interior delamination cracks also appear to diminish as the off-axis angle reduces from 90°. Many published studies in the literature report the delamination associated with the free edges of specimens. Caution must be exercised in viewing this mode of damage that results from the three-dimensional stress state existing at and within a short distance (typically one laminate thickness) from the free edges. This damage mode would thus be relevant to situations where free edges exist such as around a hole or a cutout.

6.2.3   EFFECTS OF MANUFACTURING DEFECTS

The manufacturing-induced defects in laminates lie within the plies as voids and resin-rich zones as well as fiber misalignments and fiber distribution irregularities. Additionally, dry (unbonded) regions at ply interfaces can exist. Under the axial tension of a laminate, off-axis plies are subjected to transverse tension and in-plane shear stresses. These ply stresses add to any residual stresses that are induced by the thermal cooldown of laminates during manufacturing. On imposing axial tension on a laminate, the local stress states generated at the fiber level are enhanced by the defects within the plies and can become critical for initiating cracks. The effects of defects on the initiation of such cracks have been studied under transverse tension, in-plane shear and in combined stressing, and the relevant studies were reviewed in the previous chapter (Chapter 5). The ply constraint effects on the formation of these cracks under the influence of defects have also been studied to a limited extent.

6.3   LOADING MODE: AXIAL COMPRESSION

6.3.1   CROSS-PLY LAMINATES

When a cross-ply laminate is loaded in compression in the axial direction, cracks do not form in the 90° plies. Instead, the axial 0° plies undergo microbuckling of fibers. The microbuckling phenomenon is sensitive to fiber misalignment, as reviewed in the previous chapter (Chapter 5). The constraint of the 90° plies is likely to subdue the fiber microbuckling by providing lateral support to fibers in the 0° plies. This lateral support is likely to depend on whether the 0° plies are at the surface or within the laminate thickness, i.e., whether the laminate layup is $[0_n/90_m]_s$ or $[90_m/0_n]_s$ and the relative thicknesses of 0° and 90° plies, i.e., n/m ratio. The growth of kink bands once triggered from the microbuckling will also depend on the ply constraint. Direct observations to verify these effects are, however, not easy as the final failure of the laminate under axial compression results in crushed surfaces. The final failure typically comes from out-of-plane buckling of the outer plies, which mask the precursor failure modes. The precursor mechanisms of fiber microbuckling and kink-band formation and growth in the 0° plies are likely to cause delamination by debonding the 0/90 interfaces. The ply constraint removed by delamination is then likely to trigger ply buckling.

6.3.2   GENERAL LAMINATES

In multidirectional laminates with plies in different orientations, including in 0° direction, the final failure can come from compression failure of the 0° plies, although other failure modes in the off-axis plies can occur before depending on the orientation of those plies. These failure modes are likely to be governed by shear stresses in the matrix of the plies. Additionally, delamination caused by interlaminar shear is also possible before compression failure of the fibers in the

0° plies. The concurrent and sequential occurrence of all these failure modes has not been possible to delineate in general laminates because of the complex nature of final failure that does not generally preserve the evidence of previous failure mechanisms.

Because of the difficulties of following the entire sequence of failure mechanisms and modes in general laminates under axial compression, as outlined above, most studies have supplemented the experimental observations with analytical and/ or computational simulations to come up with a description of failure evolution. A recent study [7] compared the compression failure evolution in laminates reported by different studies to show the uncertainty that currently lies in knowing the actual failure events and their sequences. The authors of that study then proposed a classification of their own to describe the reported failure modes. Their classification, shown in Figure 6.5, separates the purely fiber-dominated kink-band failure from the fiber/matrix interface-related and ply/ply interface-related failures. It also places the compression failures that have significant shear stress effects in a different class. To understand the effects of manufacturing defects on compression failure, which is the focus of this chapter, the classification displayed in Figure 6.5 can be taken to conveniently describe the effects of manufacturing defects in each of those classes. This will in this context be useful to discuss the effects of defects on kink-band formation and growth in terms of the early stages of compression failure followed by the role of defects in causing interfacial failure in the subsequent stages leading to total failure. This will be done next.

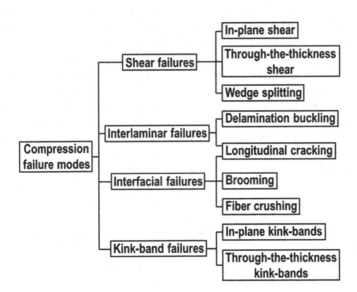

FIGURE 6.5 A proposed classification of the compression failure modes. Source: Opelt, C.V., Cândido, G.M. and Rezende, M.C., 2018. Compressive failure of fiber reinforced polymer composites–A fractographic study of the compression failure modes. *Materials Today Communications*, *15*, 218–227.

6.3.3 Effects of Manufacturing Defects

6.3.3.1 Kink-Band Failures

The formation and growth of kink bands under axial compression of UD composites were described in Chapter 5. The effect of combined loading was also discussed there. When UD layers are within a laminate, an axial compression on the laminate will generally impose a combined stress state consisting of axial and transverse normal stresses and in-plane shear stress on those layers. The formation and growth of kink bands under pair-wise combined loading, i.e., axial compression + transverse tension, and axial compression + in-plane shear, were also discussed in Chapter 5, along with the effects of fiber misalignment, fiber waviness, and matrix voids. These effects depend on the level of the stresses that are combined with the axial compressive stress which is mainly responsible for triggering fiber microbuckling leading to kink-band formation and growth. A systematic experimental study of these effects covering all possible stress combinations with respect to each type of manufacturing defect would be too extensive to conduct. The nature of the kink-band formation and growth mechanisms and the conditions favorable to their occurrence have been clarified and continue to be further clarified as more studies using XCT are conducted. Such studies aided by computational modeling and simulation are the way forward.

6.3.3.2 Interfacial Effects on Compression Failure

The role of the fiber/matrix interface in initiating kink bands was studied by comparing compression failures in UD composites where the fibers were sized, unsized, or coated with urethane [8]. Although the interface properties cannot be quantified uniquely, and can only be characterized loosely as strong, weak, brittle, tough, etc., they do have an influence on the initiation and growth of kink bands. While modeling approaches will be discussed later in the book, at this point it can be surmised that the fiber/matrix interface does have an influence on whether kink bands form and how they grow. In the context of laminates, it is important to know whether the kink bands grow in the plane, or they grow in the thickness direction of the layer where they form. This aspect is of minor consequence for the ultimate failure of a UD composite but can make a big difference in influencing the failure events after kink-band growth in laminates. This is similar to failure occurring from a single crack in a $[90]_n$ composite in transverse tension versus the final failure of $[0_n/90_m]_s$ laminate where the same 90° plies are constrained by the 0° plies. The failure events after the formation of a transverse crack in the 90° plies in this case consist of multiple transverse cracks in the 90° plies, delamination of 0/90 interface, and fiber failure in the 0° plies, as discussed above. If the same laminate is subjected to axial compression, the kink band(s) formed lead to fiber failures in the 0° plies but before the final failure of the laminate, the 0/90 interface delaminates causing out-of-plane buckling and fiber crushing noted in Figure 6.6.

The direction of kink bands in a 0° composite under axial compression, i.e., in-plane or out-of-plane, has been studied in [9]. In that study, the authors observed kink bands under axial compression of $[0]_2$ and $[0]_3$ carbon/epoxy composites made from prepregs of 0.1 mm thickness and ~65% fiber volume fraction. Small rectangular

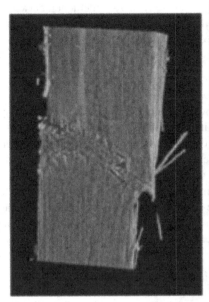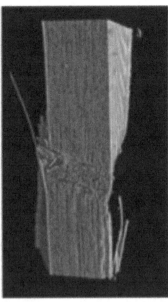

FIGURE 6.6 In-plane (left) and out-of-plane (right) kink bands observed within the volume of a UD carbon/epoxy composite under axial compression by microfocus X-ray computed tomography. Source: Ueda, M., Mimura, K. and Jeong, T.K., 2016. In situ observation of kink-band formation in a unidirectional carbon fiber reinforced plastic by X-ray computed tomography imaging. *Advanced Composite Materials*, 25(1), 31–43.

specimens 0.5 mm wide were observed in a microfocus XCT instrument. Slicing in three orthogonal planes of the composite was performed for the tomography to construct three-dimensional images of the kink bands. Figure 6.6 taken from [9] shows two typical images of in-plane and out-of-plane kink bands. The angle of inclination of the kink bands was approximately the same in both cases and failure from fiber breakage was also similar.

When kink bands form in an internal ply of a laminate, it is difficult to detect and observe those unless the bands have initiated from the edge of a specimen, or they have propagated from inside to the free edge surface. This is particularly difficult in the case of out-of-plane kink bands. Thus, recourse is often taken to modeling and simulation studies. From the point of view of failure events after kink bands, the out-of-plane kink bands are likely more important than the in-plane ones in causing delamination of the constraining plies. From observations, it is difficult to confirm this as the delamination event occurs often close to the final failure where the crushed fibers and other dynamic failure events cloud the clarity of the failure sequence.

From the point of view of the effects of manufacturing defects, there is no clear evidence of how any defects in ply/ply interfaces affect the kink band-induced delamination. It is often the case that if the outer plies of a laminate are separated by out-of-plane buckling, then the instability of this buckling is governed by structural parameters such as ply thickness and bending stiffness. There is also an interaction between local buckling and global structural buckling that is possible, which can make assessment of the effect of manufacturing defects by experimentation further difficult.

6.4 DELAMINATION

When laminates with a thermosetting polymer matrix are prepared by stacking layers in a selected layup and then cured in an autoclave, some amount of resin flows through the layers and accumulates between the layers. A relatively thin resin-rich layer that forms between the layers defines the interface between two layers and the failure of this interface is what is termed as delamination. It should be noted that in the composites literature sometimes the failure of the fiber/matrix interface is referred to as delamination. This failure should be appropriately called "debonding", while "delamination" should only be used to describe the separation of laminae at interfaces. Alternatively, delamination can be called interlaminar failure.

FIGURE 6.7   An X-ray radiograph of a carbon/epoxy laminate showing delamination from a free edge of the laminate. Source: Sun, C.T. and Zhou, S.G., 1988. Failure of quasi-isotropic composite laminates with free edges. *Journal of Reinforced Plastics and Composites*, 7(6), 515–557.

Delamination has been a critical failure mode of concern as it represents a failure of the lamination which forms the basis of generating desired properties by combining different fiber orientations. When layers of two different fiber orientations are bonded together, the interlaminar (interfacial) stresses are produced on loading the assembly of layers. These stresses can become high and exceed the capacity of the material between the layers (interphase or resin-rich layer) to sustain the stresses resulting in delamination. Stress analysis (to be discussed in a later chapter) has shown that high levels of local interlaminar stresses occur at free edges, e.g., at holes, cutouts, and notches, and at the fronts of intralaminar (matrix) cracks as these cracks approach the interlaminar regions. An illustrative example of delamination occurring at a free edge of a laminate is shown in Figure 6.7 [10]. The X-ray radiograph shown in the figure was taken after tensile loading at 22.5° to the longitudinal direction (to enhance interlaminar shear stresses) of a straight-edged specimen of carbon/epoxy of [0/90/+45/−45]$_s$ layup. An interaction of matrix cracks with delamination can also be seen in this image.

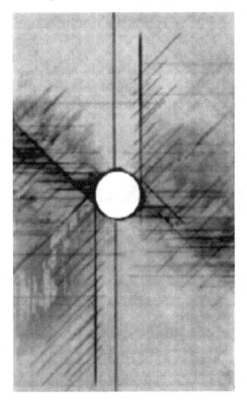

FIGURE 6.8 An X-ray radiograph of a carbon/epoxy laminate under axial tension showing delamination from the hole edge spreading out aided by matrix ply cracks. Source: Wisnom, M.R., 2012. The role of delamination in failure of fibre-reinforced composites. *Philosophical Transactions of the Royal Society A: Mathematical, Physical and Engineering Sciences,* *370*(1965), 1850–1870.

The delamination at straight free edges in laminates, e.g., at ply drop-offs, occurs, but more commonly it occurs at curved free edges such as at holes and cutouts. One example from a study [11] that looked at the effect of grouping plies of an orientation together or dispersing them in a laminate is shown in Figure 6.8 where delamination initiating from the edge of a circular hole is seen. The axial tension applied in this case to the $[45_4/90_4/-45_4/0_4]_s$ laminate produced matrix ply cracks that further enhanced delamination by inducing local interlaminar stresses.

If axial compression instead of axial tension is applied to a laminate with a hole, delamination still occurs from the hole edge but is less pronounced. An example of this is displayed in Figure 6.9 [12] for a carbon/PEEK $[(45/-45/0_2)_3]_s$ laminate where an X-ray radiograph is shown. Also here the delamination initiated at the hole edge spreads out into the laminate plane aided by intralaminar matrix cracks.

6.4.1 DELAMINATION IN CURVED LAMINATES

Use of curved laminates is not uncommon and if the curvature is low, the delamination is not significantly different from that in flat laminates. However, if corners are created by curving laminates, such as in aircraft wing and wind turbine blade spars, then the delamination from free edges and from intralaminar cracks can be severe. One study [13] has reported delamination in L-shaped laminates of carbon/epoxy cross-ply layups subjected to four-point bending to induce a constant bending moment. The inner side of the corner with tensile normal stress showed delamination

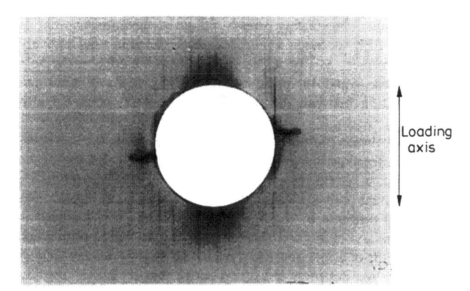

FIGURE 6.9 An X-ray radiograph of a carbon/PEEK $[(45/-45/0_2)_3]_s$ laminate under axial compression. Source: Fleck, N.A., Jelf, P.M. and Curtis, P.T., 1995. Compressive failure of laminated and woven composites. *Journal of Composites Technology & Research*, *17*(3), 212–220.

induced by matrix crack, see Figure 6.10. A main difference from flat laminates lies in the kinking of delamination cracks from one interface to a neighboring interface.

6.4.2 MANUFACTURING-INDUCED DELAMINATION

A common source of initial delamination in composite laminates is machining, e.g., drilling and cutting holes and notches in laminates. Figure 6.11 illustrates drilling-induced delamination and provides evidence of delamination around a drilled hole

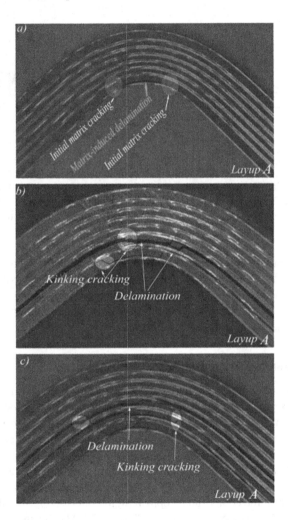

FIGURE 6.10 Delamination at the tension side of a bent corner of a cross-ply laminate. Matrix crack induced delamination and its kinking to a neighboring interface are indicated. Source: Cao, D., Hu, H., Duan, Q., Song, P. and Li, S., 2019. Experimental and three-dimensional numerical investigation of matrix cracking and delamination interaction with edge effect of curved composite laminates. *Composite Structures*, 225, 111154.

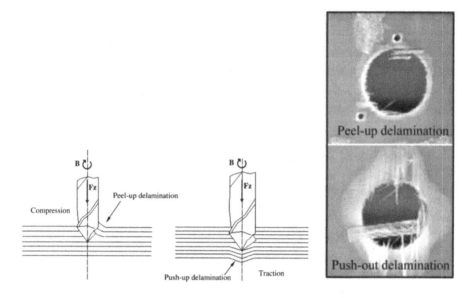

FIGURE 6.11 Schematic of the delamination at the tool entry and at the tool exit in a drilling operation on laminates, and images of the laminate surfaces (right). Source: Geng, D., Liu, Y., Shao, Z., Lu, Z., Cai, J., Li, X., Jiang, X. and Zhang, D., 2019. Delamination formation, evaluation and suppression during drilling of composite laminates: A review. *Composite Structures*, *216*, 168–186.

in a cross-ply carbon/epoxy laminate [14]. The extent of delamination was found to depend on the cutting speed and was more extensive on the tool exit side. A recent review [15] provides information on drilling-induced delamination in composite laminates where the delamination quantification methods and effects of drilling operation parameters are reviewed.

6.4.3   Delamination Due to Incorrect Molding

In an autoclave molding of laminates, the prepregs are stacked and placed in an autoclave where a specified curing process is conducted. The process consists of applying vacuum, pressure, and temperature at specified times in what is known as a curing cycle. If irregularities in conducting the curing cycle are encountered, they can result in voids, incomplete curing, and dry interfaces between the plies. One study [16] varied the pressure during autoclave molding of a carbon/epoxy UD composite while keeping other curing cycle parameters as specified and observed the cross section of the composite by optical digital microscopy and SEM. The pressure in the autoclave was varied between 0.0 and 0.6 MPa in incremental steps of 0.2 MPa. In Figure 6.12 the optical and SEM micrographs are shown at the four pressure values. As can be seen in the optical micrographs, the air trapped between the prepregs forms voids that vary from circular geometry to strip-like shape. The large strip-shaped voids become thinner at higher pressure and are no longer visible

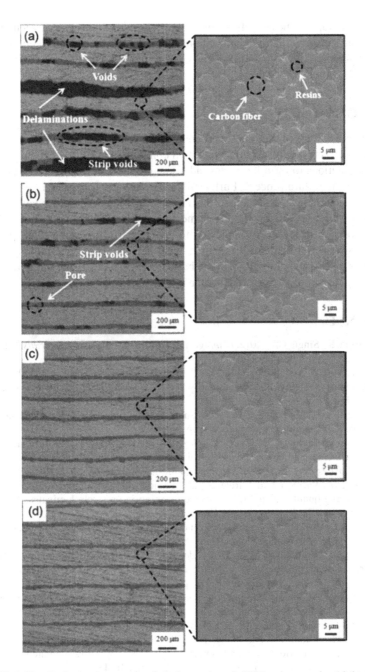

FIGURE 6.12 Optical micrographs (left images) and SEM micrographs (right images) taken on the cross section of a carbon/epoxy UD composite after autoclave molding using (a) 0.0, (b) 0.2 MPa, (c) 0.4 MPa, and (d) 0.6 MPa pressure. Source: Chang, T., Zhan, L., Tan, W. and Li, S., 2017. Effect of autoclave pressure on interfacial properties at micro-and macro-level in polymer-matrix composite laminates. *Fibers and Polymers*, *18*, 1614–1622.

in the micrographs. These voids result in dry areas in the interface between plies and can be considered as initial delaminations. Such delaminations are likely to grow if compression buckling of the plies occurs and if interlaminar stresses are induced by other loading modes.

6.5   CONCLUSION

In Chapter 5 failure in unidirectional composites in the presence of manufacturing defects was described. When unidirectional composites are stacked as layers in different orientations to form laminates, additional manufacturing defects emerge that also affect the failure process. Furthermore, the failure process in a layer within a laminate also changes due to the constraints imposed by the other layers. In this chapter, we have examined the experimental evidence regarding failure initiation and progression in laminates subjected to different loading modes with particular attention given to the effect of manufacturing defects. The observations and measurements discussed here will form the basis for approaches to the modeling of failure in laminates to be treated in Chapter 8.

REFERENCES

1. Talreja, R., Singh, C.V., 2012. *Damage and Failure of Composite Materials*. Cambridge University Press, Cambridge.
2. Garrett, K.W. and Bailey, J.E., 1977. Multiple transverse fracture in 90° cross-ply laminates of a glass fibre-reinforced polyester. *Journal of Materials Science*, *12*, pp. 157–168.
3. Talreja, R., 1985. Transverse cracking and stiffness reduction in composite laminates. *Journal of Composite Materials*, *19*, pp. 355–375.
4. Okabe, T., Imamura, H., Sato, Y., Higuchi, R., Koyanagi, J. and Talreja, R., 2015. Experimental and numerical studies of initial cracking in CFRP cross-ply laminates. *Composites Part A: Applied Science and Manufacturing*, *68*, pp. 81–89.
5. Huang, Y., Varna, J. and Talreja, R., 2014. Statistical methodology for assessing manufacturing quality related to transverse cracking in cross ply laminates. *Composites Science and Technology*, *95*, pp. 100–106.
6. Varna, J., Joffe, R., Akshantala, N.V. and Talreja, R., 1999. Damage in composite laminates with off-axis plies. *Composites Science and Technology*, *59*(14), pp. 2139–2147.
7. Opelt, C.V., Cândido, G.M. and Rezende, M.C., 2018. Compressive failure of fiber reinforced polymer composites–A fractographic study of the compression failure modes. *Materials Today Communications*, *15*, pp. 218–227.
8. Narayanan, S. and Schadler, L.S., 1999. Mechanisms of kink-band formation in graphite/epoxy composites: A micromechanical experimental study. *Composites Science and Technology*, *59*(15), pp. 2201–2213.
9. Ueda, M., Mimura, K. and Jeong, T.K., 2016. In situ observation of kink-band formation in a unidirectional carbon fiber reinforced plastic by X-ray computed tomography imaging. *Advanced Composite Materials*, *25*(1), pp. 31–43.
10. Sun, C.T. and Zhou, S.G., 1988. Failure of quasi-isotropic composite laminates with free edges. *Journal of Reinforced Plastics and Composites*, *7*(6), pp. 515–557.
11. Wisnom, M.R., 2012. The role of delamination in failure of fibre-reinforced composites. *Philosophical Transactions of the Royal Society A: Mathematical, Physical and Engineering Sciences*, *370*(1965), pp. 1850–1870.

12. Fleck, N.A., Jelf, P.M. and Curtis, P.T., 1995. Compressive failure of laminated and woven composites. *Journal of Composites Technology & Research*, *17*(3), pp. 212–220.
13. Cao, D., Hu, H., Duan, Q., Song, P. and Li, S., 2019. Experimental and three-dimensional numerical investigation of matrix cracking and delamination interaction with edge effect of curved composite laminates. *Composite Structures*, *225*, 111154.
14. Davim, J.P., Rubio, J.C. and Abrao, A.M., 2007. A novel approach based on digital image analysis to evaluate the delamination factor after drilling composite laminates. *Composites Science and Technology*, *67*(9), pp. 1939–1945.
15. Geng, D., Liu, Y., Shao, Z., Lu, Z., Cai, J., Li, X., Jiang, X. and Zhang, D., 2019. Delamination formation, evaluation and suppression during drilling of composite laminates: A review. *Composite Structures*, *216*, pp. 168–186.
16. Chang, T., Zhan, L., Tan, W. and Li, S., 2017. Effect of autoclave pressure on interfacial properties at micro-and macro-level in polymer-matrix composite laminates. *Fibers and Polymers*, *18*, pp. 1614–1622.

7 Fatigue Damage in the Presence of Defects

7.1 A CONCEPTUAL FRAMEWORK FOR FATIGUE DAMAGE INTERPRETATION

In the 1981 paper [1] this author presented a systematic characterization of fatigue damage in composite materials and proposed fatigue life diagrams (FLDs) to delineate dominant mechanisms in regions of the diagrams. Such diagrams facilitate the interpretation of the operating mechanisms underlying fatigue and clarify the roles of fiber and matrix properties as well as of ply orientations with respect to loading directions. The motivation for constructing the FLDs is to reduce the amount of testing that would otherwise be necessary to cover all possible combinations of loading and material parameter variations. To start, the baseline case of fatigue under tension–tension loading in the fiber direction is treated by addressing the basic nature of fatigue, i.e., the failure event(s) occurring in the first application of a load, the accumulation of irreversible mechanisms in subsequent load applications, and the attainment of a critical state that causes final failure. These considerations led to the construction of the baseline FLD as shown in Figure 7.1. As seen, the maximum axial strain ε_{max} attained in the first application of load (in a load-controlled test) is plotted against the number of cycles N (on a logarithm scale). Two characteristic values of the UD composite, i.e., the failure strain ε_c, and the fatigue limit ε_{fl} are marked on the vertical axis. Between these two strain values, the FLD is depicted in two regions, Region I representing fiber-dominated mechanisms and Region II where matrix and fiber-matrix interface play dominant roles. Below the fatigue limit ε_{fl} is Region III where irreversible mechanisms either do not initiate or, if initiated, do not accumulate to the critical state of final failure. The role of fiber stiffness is indicated in delaying the damage accumulation in Region II and thereby enhancing the fatigue limit.

The scatter band, indicative of the randomness in static failure is shown at the failure strain ε_c, and similarly, the randomness in damage accumulation resulting in variability of fatigue life is shown by the scatter band in Region II of the FLD. The flatness of the Region I scatter band is an important novel feature of the FLD, which is often misunderstood and/or ignored in the literature and will be described next.

As shown in the FLD in Figure 7.1, the dominant mechanism in Region I is assumed to be independent of load cycles resulting in the horizontal appearance of the scatter band in this region. The scatter band indicates randomness in the static failure, which is due to the inherent randomness of fiber failure. The placement of the scatter band without dependency on the load cycles suggests that damage

DOI: 10.1201/9781003225737-7

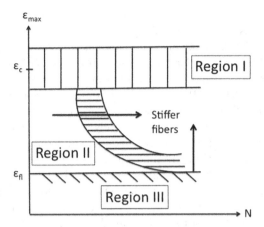

FIGURE 7.1 The baseline fatigue life diagram (FLD) for tension-tension loading in the fiber direction of UD composites shows three regions of the diagram. The vertical axis indicates the maximum strain reached in the first load cycle in a load-controlled test.

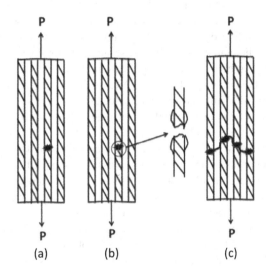

FIGURE 7.2 Fiber failure scenarios under a tensile load P. (a) A dry fiber bundle with the weakest fiber failure, (b) A fiber composite with the weakest fiber failure and debonding over a short length, and (c) Composite failure from linkage of failed fiber regions.

accumulation, if any, has a negligible effect on fiber breakage. This is explained by failure scenarios sketched in Figure 7.2. Consider a dry bundle of fibers loaded by a tensile force P, as illustrated in Figure 7.2(a). As is known, fibers produced commercially have defects resulting in their tensile strength over a given length being determined by the weakest point in that length. Thus, in a bundle of a given length where all fibers are parallel and of equal length and diameter, the fiber (or fibers) with the weakest point will fail first as the bundle is subjected to a tensile load P. Let $P = P_1$

be the value at which the first fiber failure occurs. At this load, the surviving fibers in the bundle will instantaneously and equally share the load released by the broken fiber(s). If none of the surviving fibers fail at the new stress value corresponding to the applied load P_1, then on unloading and reloading the bundle to that load no new fibers will fail. In fact, assuming the fibers have no time-dependent failure process, any reapplication of the load P_1, any number of times, will not cause new fiber failures. In other words, there will be no fatigue failure of the fiber bundle.

Consider now a composite consisting of a polymer matrix reinforced by the same bundle and assume that all fibers in the bundle are perfectly bonded to the matrix. The axial tensile load P applied to this composite will at some load value $P = P_2$ cause the same fiber (or fibers) to fail first that failed also first in the dry bundle. However, the consequence of the fiber failure(s) in the composite will be quite different from that in the dry bundle case. First, only the fibers in the immediate neighborhood of a broken fiber will carry the additional stress released by the broken fiber. Second, the broken fiber will debond locally from the matrix, as illustrated in the zoomed-in sketch next to the composite, Figure 7.2(b). Third, the matrix surrounding the broken fiber ends will be stressed at a significantly higher level than prior to the fiber failure. Most importantly, at this new stress level, the matrix polymer will likely deform inelastically, producing the irreversibility that is the key to the fatigue process. Thus, unloading the load from P_2, and reloading to this value will change the stress states in the matrix as well as in the surrounding fibers in the close neighborhood to the broken fiber. Repeated application of this load can result in an accumulative process that can cause fatigue failure of the composite if a critical failure condition for composite failure is reached. A scenario for such criticality is sketched to the right in Figure 7.2 indicating failure from the linkage of broken fibers in a local neighborhood of the weakest fiber.

In a UD composite of a given volume, the fiber defects will be distributed in the volume depending on the manufacturing process. Consequently, under axial tensile load, the volume-averaged failure stress (or strain) will not be deterministic but will have a probability distribution. Since the stresses in the fibers and in the matrix are not the same, for failure value we will refer to the average strain in the composite, ε_c, as this is also the nominal fiber failure strain irrespective of the fiber volume fraction. The scatter in the failure strain as the result of the randomness of the failure process is described by certain convenient extreme values such as those at 5% and 95% probabilities of failure.

Now let the composite be subjected to an axial tensile load such that the resulting maximum strain, ε_{max}, is within the scatter band of the composite failure strain, ε_c, shown in Figure 7.1. If the composite survives this load, then it can be concluded that none of the fiber failure regions reached the failure condition discussed above and illustrated in Figure 7.2(c). For this condition to occur, sufficiently many fibers must fail in a local region such that the matrix crack formed grows unstably at the maximum applied load. Since the load applied is high enough for the maximum strain to be within the failure scatter band, it is reasonable to assume that many fibers fail in different regions at this load, although the failure condition is not reached in any of those regions. Consider now unloading and reloading to the same maximum load value. In the reapplication of load, each of the fiber failure regions will undergo stress redistribution due to the inelastic deformation of the matrix that has occurred before. However, the resulting stress fields in the regions will be different and their consequence in terms of failing more fibers will also be different because of the random distribution of defects (weak points) in the fibers. If the reapplication of the maximum load is repeated, then with each repetition a new scenario of the fiber failure regions will form with the region most likely to fail the composite in the next application of the maximum load shifting from region to region. Thus, which of the failure regions will reach the composite failure condition first cannot be predicted since a progressive mechanism that depends on the maximum applied load does not exist. Consequently, the composite failure is not a function of the number of load cycles N and the probability of this failure is also independent of N. In other words, the scatter band of initial failure strain remains unchanged with the number of load cycles N.

To understand the construction of Region II of the FLD consider now loading the UD composite in tension such that the maximum strain in the first application of load is significantly below the scatter band of ε_c shown in Figure 7.1. Under this load, few if any fibers will be expected to break. However, a real possibility exists that some fibers are broken due to manufacturing, e.g., from handling the fibers in dry form or stretching them during impregnation with the matrix polymer. The broken fiber ends are sites of stress concentration in the matrix and, consequently, excursion of the matrix deformation into the inelastic regime. Thus, under repeated load application, these sites will be sources of irreversibility needed to initiate and advance the fatigue cracks. The role of fibers now will be to delay, and possibly arrest, the growth of these fatigue cracks. If the repeating applied load is high enough, then the matrix cracks will grow as fiber-bridged cracks, and the crack with the largest cyclic growth rate will reach the composite failure condition. That failure condition can be expressed in energy terms as exceeding the local fracture resistance, i.e., by the energy release rate of a fiber-bridged matrix crack exceeding a critical material value. Satisfaction of this criticality condition depends on the size of the crack and the remotely applied load level. As this load level is reduced, the crack growth rate will reduce, and consequently, the number of load repetitions needed to reach the failure condition will increase. This trend results in the fatigue life scatter band sloping downwards, as depicted in Figure 7.3. The scatter band represents the statistical variation caused by randomness in the crack growth rate as well as in the local fracture toughness at the crack fronts. The fiber-bridged matrix crack as the dominant mechanism is also depicted in Figure 7.3.

In many applications, one is concerned with the threshold load below which fatigue failure will not occur in many load cycles, typically 10^6 or higher. This condition is given by the fatigue limit strain ε_{fl}, below which the fiber-bridged matrix crack of Region II will not reach unstable crack growth. A conservative estimate of this strain can be the fatigue limit of the matrix polymer measured under strain-controlled loading. At that strain, the matrix will not develop a crack with the potential to grow unstably and will likely therefore not be bridged by fibers.

The baseline FLD for UD composites described here and its variations to address fatigue in UD composites under combined loading and in laminates of different ply

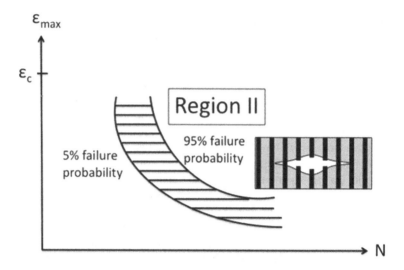

FIGURE 7.3   Region II of the baseline fatigue life diagram.

stacking has been described in [2–11]. The purpose here is to use the FLDs as a guide to review the experimental observations of the fatigue process in composite materials. In particular, the interest is to gain insight into the effects of manufacturing defects on the composite fatigue damage to assess structural performance by modeling to be treated in Chapter 9.

7.2   OBSERVATIONS OF FATIGUE DAMAGE IN UD COMPOSITES

7.2.1   AXIAL TENSION–TENSION LOADING

Fatigue damage mechanisms in UD composites of a carbon/epoxy under tension–tension loading were studied and their interpretation using the FLD was reported in [12]. The fatigue mechanisms were observed at intermittent load cycles by a surface replication technique where a cellulose acetate tape made malleable with an acetone solution was pressed against the specimen surface and hardened. Micrographs of the replicas provided images of fatigue damage mechanisms. Some of the key observations are described here. Figure 7.4 displays evidence of a fiber-bridged crack stipulated in Region II (Figure 7.3). The crack captured in the first cycle was found to grow past a few fibers resulting in its opening profile changing from an unconstrained one (as predicted by fracture mechanics) to the crack tips getting squeezed together by the bridging fibers. An expanded image of a crack shown in Figure 7.5 displays the fiber bridging effect more clearly.

Observations in Region III are illustrated in Figure 7.6 where one sees a matrix crack emanating from a fiber break being arrested after initial growth. The crack-arresting mechanism was found to be caused by the diversion of the matrix crack fronts into fiber-matrix interfaces. The subsequent cyclic growth of the debond crack was found to cease after a certain distance from the matrix crack tip.

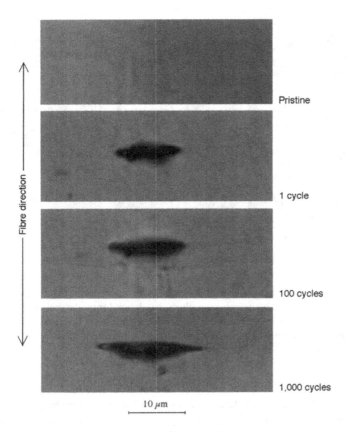

FIGURE 7.4 Fiber-bridged crack propagation in Region II of the FLD for carbon/epoxy. Source: Gamstedt, E.K. and Talreja, R., 1999. Fatigue damage mechanisms in unidirectional carbon-fibre-reinforced plastics. *Journal of Materials Science*, *34*, 2535–2546.

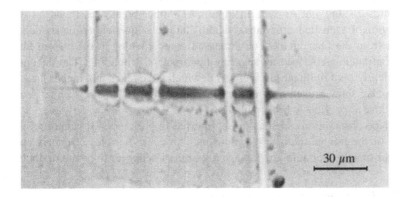

FIGURE 7.5 A fiber-bridged crack of Figure 7.4 showing the bridging effect more clearly. Source: Gamstedt, E.K. and Talreja, R., 1999. Fatigue damage mechanisms in unidirectional carbon-fibre-reinforced plastics. *Journal of Materials Science*, *34*, 2535–2546.

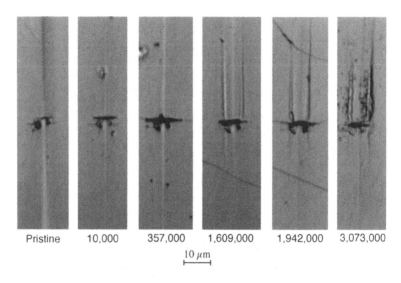

Pristine   10,000   357,000   1,609,000   1,942,000   3,073,000

10 μm

FIGURE 7.6   A crack emanating from a broken fiber unable to grow in Region III loading because of being arrested by neighboring fibers that debond from the matrix if the loading is continued over a very large number of cycles. Source: Gamstedt, E.K. and Talreja, R., 1999. Fatigue damage mechanisms in unidirectional carbon-fibre-reinforced plastics. *Journal of Materials Science*, *34*, 2535–2546.

Another study [13] of fatigue under axial tension-tension loading on a glass/epoxy UD composite focused on the fiber breakage process. As was already noted in [1], for composites of low-stiffness fibers such as glass, Region I of the FLD extends only to a few hundred load cycles. Thus, most of the fatigue damage corresponds to Region II. The authors of [13] made optical microscopy observations on a polished surface region of the specimens. Images were taken intermittently at different number of load cycles and at different load levels. Some of the key observations are reproduced in Figures 7.7 and 7.8. As seen in Figure 7.7 a fiber break was found on the first application of a tensile load. Matrix yielding and subsequent fiber-matrix debonding ensued from the fiber break. The increased crack opening at the broken fiber end with continued cyclic loading suggests fiber-matrix debonding. The debonding is more clearly seen in images in Figure 7.8, which were taken at a higher load. These images also show broken fiber cluster formation likely from stress enhancement in fibers close to previously broken fiber(s).

Surface observations such as those reported in [12] and [13] cannot provide all details of the fatigue damage mechanisms. However, the roles of matrix cracking and fiber-matrix debonding in damage accumulation appear to be not in doubt.

7.2.2 TRANSVERSE TENSION–TENSION LOADING

Failure of UD composites under cyclic transverse tension has been of major interest because in laminates the first failure event often occurs in tension normal to fibers in

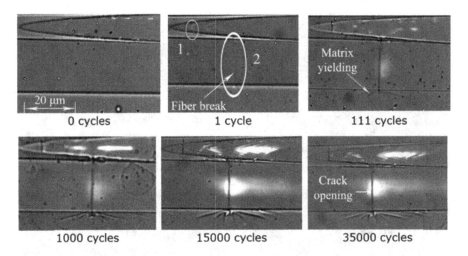

FIGURE 7.7 Fiber break and ensuing damage in a glass/epoxy UD composite in tension-tension fatigue. Maximum stress 300 MPa. Source: Castro, O., Carraro, P.A., Maragoni, L. and Quaresimin, M., 2019. Fatigue damage evolution in unidirectional glass/epoxy composites under a cyclic load. *Polymer Testing*, *74*, 216–224.

off-axis plies. The formation of this crack was treated in Chapter 5 under static loading. Here we address the same failure mechanism under cyclic transverse tension.

Historically, in the late 1970s the transverse cracks in 90° plies of cross ply laminates were observed in glass/polyester composites and they drew a lot of attention as they revealed the nature of multiple ply cracking [14]. It was observed that failure in cross ply laminates was not occurring from a single crack, as in metals, but more cracks were forming whose mutual spacing in the transverse plies was reducing and tending to a uniform "saturation" crack spacing. These observations prompted studies aimed at predicting the crack spacing evolution as a function of the applied tensile load. More experimental studies were conducted that clarified the effect of transverse ply thickness on the multiple cracking and the initiation strain for first crack formation (later coined as "first ply failure"). One of the issues of engineering importance has been to find how the first crack forms. The observation tools in the early studies were not adequate to provide clear details of the crack-forming process but from optical microscopy it was inferred that fiber-matrix debonding was the first event in transverse crack formation [15]. Later in a comparative study of monotonic versus cyclic stressing of cross ply laminates [16] the authors found that in fatigue the formation of transverse cracks is at a lower strain. They noted that for thin transverse plies the fiber-matrix debond cracks did not develop into full (across the thickness) transverse cracks, while they did so for thicker plies. A more recent study produced clear evidence of the transverse crack formation under cyclic stressing [17].

The evidence of transverse crack formation shown in Figure 7.9 [17] is often cited in the literature because of its clarity and is taken also to illustrate the crack formation process in monotonic stressing. This evidence has supported modeling efforts

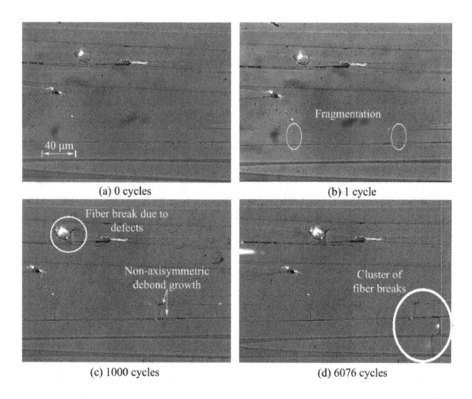

(a) 0 cycles   (b) 1 cycle

(c) 1000 cycles   (d) 6076 cycles

FIGURE 7.8   Fiber break, fiber/matrix debonding, and fiber break cluster in a glass/epoxy UD composite in tension–tension fatigue. Maximum stress 320 MPa. Source: Castro, O., Carraro, P.A., Maragoni, L. and Quaresimin, M., 2019. Fatigue damage evolution in unidirectional glass/epoxy composites under a cyclic load. *Polymer Testing*, *74*, 216–224.

aimed at understanding the influence of the local conditions such as the presence of fiber clusters and resin-rich areas on the transverse crack formation. These efforts will be reviewed later in Chapter 9 on modeling where the effect of manufacturing defects on the crack formation process will also be addressed.

7.2.3   COMBINED TRANSVERSE TENSION AND IN-PLANE SHEAR

When a laminate is subjected to cyclic axial loading, the plies with fiber orientations between the axial and transverse directions undergo fatigue under the combined effect of biaxial stresses. A review of the literature on multiaxial fatigue [18] indicated clearly that a significant interactive effect of the combined transverse tension and in-plane shear stress existed. The operative damage mechanisms under this combined loading were however not clear enough to support the modeling efforts. Since then, focused efforts to clarify these mechanisms have been conducted beginning with [19]. In [19] the authors used tubular specimens of glass/epoxy with fibers oriented in the circumferential direction. These specimens were loaded under in-phase cyclic axial tension and torsion, both loading modes with an R-ratio of 0.05. A constant biaxiality

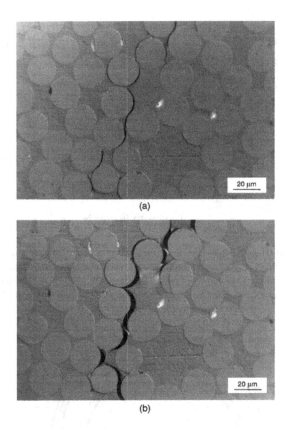

(a)

(b)

FIGURE 7.9 Fiber/matrix debonding seen in (a) develops into a continuous transverse crack as seen in (b). Source: Gamstedt, E.K. and Sjögren, B.A., 1999. Micromechanisms in tension-compression fatigue of composite laminates containing transverse plies. *Composites Science and Technology*, 59(2), 167–178.

ratio, defined as $\lambda_{12} = s_2/s_6$, where s_2 and s_6 are the amplitudes of the transverse stress and in-plane shear stress, respectively, was maintained. The test results are shown in Figure 7.10 where the fatigue life with transverse stress only is compared with that under two biaxiality ratios. A vertical line drawn at 10^6 cycles emphasizes the significant effect of the shear stress in reducing the transverse stress to failure. Visual observations aided by an infrared camera did not show a damage accumulation process; the failure was found to occur instantly at the initiation of a crack along fibers.

The plot in Figure 7.10 is in accordance with the FLD framework described above. It shows only Region II representing the progressive damage, although the visual observations were not able to find the damage accumulation process. Most of the damage lies in the initiation of a crack, the cyclic crack growth occupying a small portion of the fatigue life. The presence of the shear stress in addition to the transverse tension enhances the crack initiation process. The authors in [20] conducted the combined tension–torsion fatigue tests this time by adding a fabric layer on each

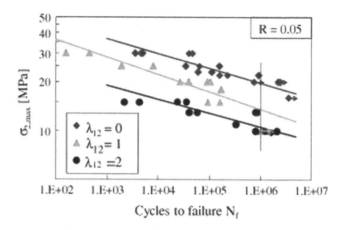

FIGURE 7.10 Maximum transverse stress on a UD composite plotted against the number of cycles to failure with and without a simultaneous cyclic in-plane shear stress of two different biaxiality ratios. Source: Quaresimin, M. and Carraro, P.A., 2013. On the investigation of the biaxial fatigue behaviour of unidirectional composites. *Composites Part B: Engineering, 54,* 200–208.

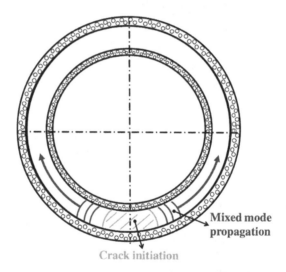

FIGURE 7.11 Schematic representation of crack initiation and mixed-mode crack propagation under cyclic tension-torsion of cylindrical tubes with fibers in the circumferential direction having layers of a glass fabric on the inner and outer sides of the tube. Source: Quaresimin, M. and Carraro, P.A., 2014. Damage initiation and evolution in glass/epoxy tubes subjected to combined tension–torsion fatigue loading. *International Journal of Fatigue, 63,* 25–35.

side of the 90° composite to slow down the crack propagation phase induced by the constraint of the fabric layers. Their observations of the crack growth process are summarized in Figure 7.11. These observations agree with the conjectured fatigue damage mechanisms in [1] reproduced here in Figure 7.12. The corresponding FLDs

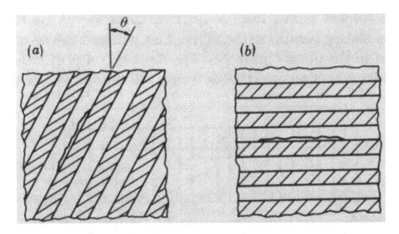

FIGURE 7.12 Matrix and interfacial cracking under off-axis fatigue of UD composites. (a) Mixed opening and sliding mode crack growth, $0 < \theta < 90°$, (b) opening mode crack growth (transverse fiber debonding, $\theta = 90°$. Source: Talreja, R., 1981. Fatigue of composite materials: damage mechanisms and fatigue-life diagrams. *Proceedings of the Royal Society of London. A. Mathematical and Physical Sciences*, 378(1775), 461–475.

shown in Figure 7.13 [1] based on these conjectures also predict the fatigue life trend in the plot of experimental data displayed in Figure 7.10 when viewed in terms of the stress (or strain) in the principal material directions of the UD composite. While the authors in [19] did not determine the fatigue limit under combined loading, the FLDs in Figure 7.13 suggest the fatigue limits (in strain) to lie between the two extremes for $\theta = 0°$ (ε_m) and for $\theta = 90°$ (ε_{db}). For glass/epoxy composites these limits have been found by testing to be $\varepsilon_m = 0.6\%$ and $\varepsilon_{db} = 0.12\%$ [1].

The role of in-plane (axial) shear may be to induce inelastic deformation in the inter-fiber region of a UD composite, but it is difficult to confirm this by in situ observations. The fracture surface observations have indicated the presence of hackles (cusps) suggesting shear-induced deformation in the failure process. Figure 7.14 taken from [20] shows SEM images of fracture surfaces at three different combined loading levels compared to transverse tension only. At low biaxiality ratio λ_{12} the fracture surface is smooth and compares with that for transverse stress only. At increasing value of this ratio, more shear cusps are seen. Such observations were also reported earlier in [21].

7.2.4 AXIAL CYCLIC COMPRESSION LOADING

Damage mechanisms in UD composites under axial compression were described in Chapter 5 where the role of fiber waviness/misalignment was also discussed. As discussed there, the formation of kink bands is tied to fiber misalignment. Once a kink band forms, its growth is governed by the inelastic deformation of the matrix. Under cyclic compression, it is expected that the hysteresis of the irreversible deformation will generate the driving force for cyclic growth and culminate in the critical

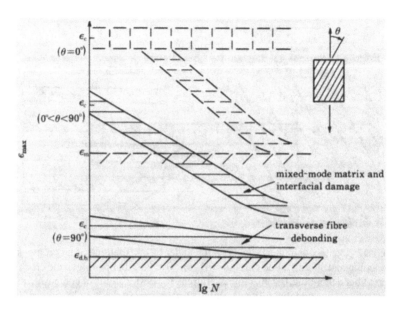

FIGURE 7.13 Fatigue life diagram for off-axis fatigue of UD composites. The dotted lines correspond to the baseline FLD ($\theta = 0°$) described above (Figure 7.1). Source: Talreja, R., 1981. Fatigue of composite materials: damage mechanisms and fatigue-life diagrams. *Proceedings of the Royal Society of London. A. Mathematical and Physical Sciences*, 378(1775), 461–475.

condition of failure. These aspects can be treated by proper models which will be discussed in Chapter 9.

From the FLD perspective, the presence of Region I is not expected as conditions for progressive damage exist under the fiber-dominated failure mechanism of kink band formation. Region II which has its upper limit at the compression failure strain will extend with a sloping scatter band to a fatigue limit. The fatigue limit strain is given by the threshold for the formation of kink bands and will be sensitive to the degree of fiber misalignment.

7.3 FATIGUE IN LAMINATES

7.3.1 CROSS PLY LAMINATES

The transverse crack formation mechanism described above for UD composites under cyclic transverse tension applies also to transverse plies in cross ply laminates with the added effect of the constraint exerted by the neighboring plies. This constraint effect depends on the thickness of the ply under consideration and the relative stiffness of constraining plies, as described in Chapter 6 for static loading. If instead of a monotonically increasing tensile load a cyclic tensile load is applied in the axial direction, then depending on the maximum load applied in the first load cycle, multiple cracks of a certain crack density are seen. This crack density increases with load cycles as new cracks between the pre-existing cracks appear and as previously

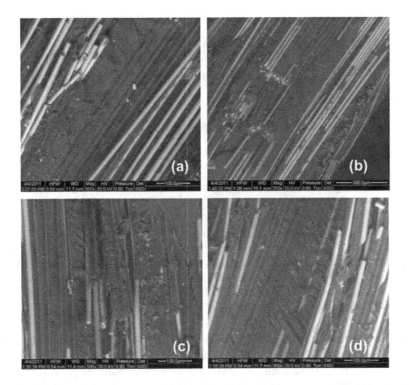

FIGURE 7.14   SEM images of fracture surfaces of tubes failed under combined cyclic normal stress and axial shear at different biaxiality ratios, $\lambda_{12} = 0$ (a), 0.5 (b), 1.0 (c), and 2.0 (d) [20]. Source: Quaresimin, M. and Carraro, P.A., 2014. Damage initiation and evolution in glass/epoxy tubes subjected to combined tension–torsion fatigue loading. *International Journal of Fatigue*, *63*, 25–35.

initiated partial cracks grow to span the ply thickness and width. Other cracks close to the ply interfaces extending in the axial direction are also seen. These cracks, often described as axial splits, are attributed to the local Poisson's contraction effect. At the intersection of these cracks with the transverse ply cracks local failure of the 0/90 ply interface occurs. This failure is termed interior delamination to distinguish it from free-edge-related delamination, which is the consequence of through-thickness stresses acting at free edges. The interior delaminations are revealed by X-ray radiography images, as shown in Figure 7.15, which also depicts the transverse cracks and axial splits in a carbon/epoxy cross ply laminate under cyclic tension [22]. It is noted that cracking modes other than the transverse ply cracks are not easily observed under a monotonic tensile load as they are usually too small and are missed even in an X-ray radiograph. Under cyclic loading, however, they can grow to sizes such that they can be detected and imaged.

Based on the observed mechanisms in cross ply laminates it is easy to construct a FLD for this case starting from the baseline FLD described above (Figure 7.1). Region I will now stand for the fiber-dominated failure mechanisms in the $0°$ plies,

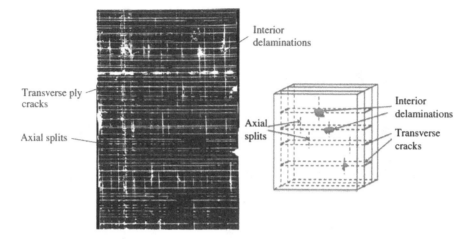

FIGURE 7.15 X-ray radiograph of a carbon/epoxy cross ply laminate subjected to tension–tension cyclic loading. The sketch to the right shows a three-dimensional depiction for clarity. Source: Reifsnider, K.L. and Jamison, R., 1982. Fracture of fatigue-loaded composite laminates. *International Journal of Fatigue*, *4*(4), 187–197.

while the progressive damage mechanisms illustrated by Figure 7.15 will be represented by Region II. The threshold beyond which these progressive mechanisms are operative, i.e., the fatigue limit, is the strain that initiates multiple ply cracking, ε_{mc}. Thus, Region III of non-evolving or un-initiated damage mechanisms will fall below this strain. Figure 7.16 depicts the FLD for cross ply laminates where the actual fatigue life data reported in [23] have been inserted.

7.3.2 General Laminates

Combining plies of different fiber orientations in stacking sequences and ply thicknesses offers numerous possibilities to create laminates for desired performance requirements. While this is an attractive feature of composite laminates in theory, in practice the problem of finding an optimal laminate configuration for fatigue performance appears to be intractable. The approach has been to seek a suitable laminate configuration within a class of laminates such as angle ply laminates ($[\pm\theta]_s$) and quasi-isotropic laminates, e.g., $[0/\pm\theta/90]_s$. These two classes of laminates will be discussed next.

7.3.2.1 Angle Ply Laminates

Based on the observations of damage in cross ply laminates it can be expected that the two off-axis plies $+\theta$ and $-\theta$ will each suffer multiple cracking under mutual constraint when a cyclic tensile load is applied in the axial direction. Also, at the fronts of the ply cracks the ply interfaces will crack locally forming strips of delaminations, as illustrated in Figure 7.17. The growth of these delaminations and coalescence into neighboring delaminations will result in the separation of the plies and consequent

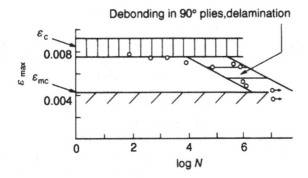

FIGURE 7.16 FLD of cross ply laminates. Source of data inserted in the figure: Grimes, G.C., 1977, January. Structural design significance of tension-tension fatigue data on composites. In *Composite Materials: Testing and Design (Fourth Conference)*. ASTM International.

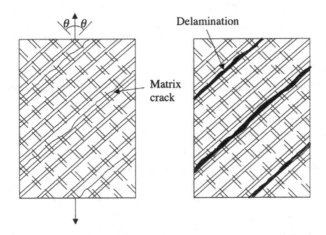

FIGURE 7.17 Multiple matrix cracks in the off-axis ply of a laminate (left) and subsequent delamination caused by fatigue (right). Cracks and delamination are shown for one ply only for clarity; the other ply is indicated in broken lines.

laminate failure. To capture these damage mechanisms in a FLD we must examine the three regions of the baseline FLD one by one. Region I, which represents fiber-dominated failure, can be expected for low values of θ. As this angle increases, the progressive mechanisms of ply cracking and interface failure are expected to start in the first application of the axial tensile load, resulting in Region I not being present. Then, Region II will start at the composite failure strain ε_c and extend down to the fatigue limit. The strain corresponding to the fatigue limit will be the strain that initiates matrix cracking in the off-axis plies on the first application of the load.

While the delamination caused by the off-axis cracks plays an increasing role as the applied load level increases, at low load levels approaching the fatigue limit the off-axis cracks will occupy a significant part of the fatigue life. The initiation of off-axis cracks is difficult to capture in an unconstrained UD composite because of the abrupt failure of a single crack. In a novel test where local axial shear was introduced by cutting fibers [24] cracks were found to initiate at an angle to the fiber axis (Figure 7.18). The local stress produced this way is likely not pure shear and judging from the orientation of cracks the maximum principal stress is not at 45° to the fiber axis. The maximum principal stress and its direction are also likely to change in the inter-fiber region. The sigmoidal shape of the cracks seen in Figure 7.18 suggests that the crack planes follow the direction normal to the local maximum principal stress and turn into the fiber/matrix interface. In fatigue cycling these cracks are likely to merge at the fiber/matrix interfaces forming a continuous crack along fibers.

The early stages of off-axis crack formation in angle ply laminates have been studied by constraining the off-axis plies by 0° plies such as in [±45/0]$_s$ glass/epoxy laminate configuration used in [25]. Figure 7.19 shows the images reported in [25] where the microcracks seen in the top 45° ply are shown. The direction of the plane normal to the local maximum principal stress is indicated in the images. On further loading, these multiple cracks in the inter-fiber region were found to merge, like those shown in Figure 7.18, to form continuous off-axis cracks. The growth of the off-axis cracks in the thickness direction of the plies was also documented in [25].

7.3.2.2   Quasi-Isotropic Laminates

This class of laminates produces directionally independent in-plane elastic properties, a desirable feature in some applications. Common examples of such laminates are [0/±60]$_s$ and [0/±45/90]$_s$. The latter is more desirable as it allows a good balance of extensional and shear properties. In terms of damage mechanisms, the presence of 0° plies in both cases produces fiber dominated non-progressive mechanism resulting in Region I of the FLD. The scatter band of Region II begins below that of Region I and extends down to the fatigue limit. The strain at the fatigue limit is the threshold for initiating the first ply cracking in the laminate. Thus, for the [0/±60]$_s$ laminate it will be the fatigue limit found for the [±60]$_s$ angle ply laminate, while for the [0/±45/90]$_s$ laminate it will be the first ply cracking strain for a cross ply laminate of the same 90° ply thickness. With these considerations, an FLD is constructed for the [0/±45/90]$_s$ laminate of carbon/epoxy and is shown in Figure 7.20 with the data

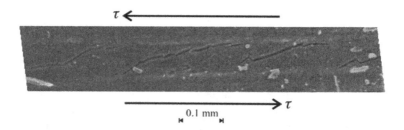

FIGURE 7.18   Cracking between fibers induced by local shear. Source: Redon, O., Fatigue damage development and failure in unidirectional and angle-ply glass fibre/carbon fibre hybrid laminates, Technical Report Risø-R-1168, 2000, Risø National Laboratory, Roskilde, Denmark.

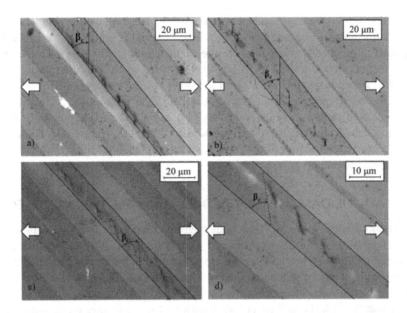

FIGURE 7.19 Examples of microcracks observed in the top 45° ply of [±45/0]$_s$ glass/epoxy laminate. Arrows indicate the loading direction. Source: Quaresimin, M., Carraro, P.A. and Maragoni, L., 2016. Early stage damage in off-axis plies under fatigue loading. *Composites Science and Technology, 128,* 147–154.

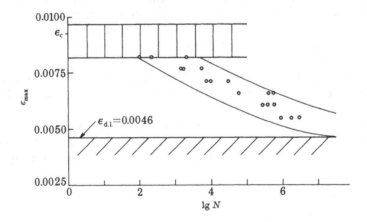

FIGURE 7.20 A FLD of a quasi-isotropic laminate with 0°, 45°, and 90° plies with data inserted from [26]. The fatigue limit strain ε_{dl} is the strain at which the transverse cracks in the 90° plies cause delamination. It is also the strain at which multiple transverse cracking occurs. Source of data inserted in the figure: Ryder, J.T. and Walker, E.K., 1977. Effect of compression on fatigue properties of quasi-isotropic graphite/epoxy composite. in *Fatigue of Filamentary Composite Materials*, STP 636, American Society for Testing and Materials, Philadelphia, 3–26.

reported in [26]. As seen in the FLD in Figure 7.20, Region I and Region III are the same as for a cross ply laminate of the same material (Figure 7.16), while where Region II falls depends on the off-axis plies of the laminate. By these considerations one can generalize the FLD to $[0/\pm\theta/90]_s$ laminates where only the placement of Region II depends on the angle θ, and the other two regions of the FLD are fixed by the 0° and 90° plies as in a cross ply laminate. The value of the FLDs can now be appreciated in that these diagrams not only allow logical interpretations of the damage mechanisms but also reduce the burden of fatigue testing that can otherwise become exorbitant because of the huge possible combinations of the laminate configurational parameters [27].

7.4　FATIGUE DAMAGE INDUCED BY MANUFACTURING DEFECTS

The previous sections treated fatigue damage in composite materials without specific reference to manufacturing defects. This is necessary for first creating a big picture of the damage mechanisms before examining the roles of defects. As a general principle, it can be stated that the presence of defects alters the local stress field which in turn influences the occurrence of failure events. It seems logical then to expect that the defects will affect the initiation of the first failure event and likely also will play a significant role in the early stages of the damage process. Once the characteristic sizes of the failure entities (e.g., fiber/matrix debonds, matrix cracks, kink bands, and ply/ply interface failure) grow beyond the governing sizes of the defects (e.g., matrix void diameter, fiber clusters, and fiber waviness), the defects may become less important as the failure entities themselves may interact to affect further evolution of damage. While such considerations are important in planning studies to explore the effects of defects, it must be realized that experimental observations/measurements have limitations and should be aided by computational simulations at different length scales. With this in mind, we review the experimental observations and in a later chapter treat the theoretical and computational studies. This is believed to be a more objective approach than analyzing the experimental data with a preselected modeling strategy.

In the following, the effects of defects on fatigue damage will be assessed in the same order as above, i.e., UD composites under different cyclic loading modes followed by composite laminates of different configurations under those loading modes.

7.4.1　UD COMPOSITES

In Chapter 5 the damage initiation, propagation, and criticality under static loading were described. Each elementary loading mode and all possible pairwise in-plane loading combinations were considered. While generally triaxial imposed loading occurs on structures, experimental observations of damage are limited to biaxial loading by the capabilities of testing machines and the specimen geometries that can be tested as well as by observation and measurement techniques. In fatigue loading these limitations are severe as discussed in [19]. In view of this, a full picture of how manufacturing defects alter/enhance the fatigue damage mechanisms in UD composites does not exist. In the following, we shall review what seems to be known and point to what remains unclear in terms of fatigue damage mechanisms. In doing this we shall follow the systematic framework of FLDs for interpretations of these mechanisms.

7.4.1.1 Axial Tension–Tension Loading

Following the baseline FLD (Figure 7.1), Region I represents statistical fiber failure that has negligible progressive damage. As described in Chapter 5, the statistical fiber failure reflects the manufacturing defects (variability) such as non-uniformity of fiber diameter and sizing on the fiber surface. These defects play roles in creating fiber failures during the first application of the tensile load. If the composite does not fail in the first cycle, then failure in subsequent load cycles is due to a small inelasticity in the matrix around the fiber breaks. This random but non-progressive process does not have a measurable effect of matrix defects, but it can be affected by fiber/matrix interfacial defects. However, unless these defects are substantial, their effect has not been found. In a significant presence of such defects the scatter band of Region I will be expected to slope downwards reflecting the progressiveness of the interfacial debond growth in fatigue.

The progressive damage in Region II of the FLD is governed by the fatigue cracking of the matrix modified by the bridging fibers. The initiation of a matrix crack before bridging can be from a broken fiber end or by cyclic inelastic deformation of the matrix between fibers. The voids in the matrix can influence this process but its documentation is not easy because of the random nature of fiber failures. In a study [28], where high-resolution computed tomography was performed, a first-order correlation between void volume fraction and fiber break density was not found but a correlation between the location of fiber breaks and voids could be inferred. Also, evidence of intersections between voids and matrix cracks was seen. This suggests that the voids can influence the progressive mechanism of Region II both by aiding fiber breaks and by initiating matrix cracking. However, a systematic study to quantify this influence is not available. This further calls for modeling and simulation efforts that will be reviewed in a later chapter.

It is argued that any defects at the fiber/matrix interfaces will also affect the progressive damage mechanisms of Region II of the FLD. Such defects are difficult to control in a fatigue damage study as they result from the manufacturing process which will need to be artificially changed from the prescribed one to allow a parametric investigation. Instead, a comparative study was conducted where the fiber/matrix interface was changed by using two polymer matrices with the same fiber. In [29] a polypropylene (PP) and a modified PP with maleic anhydride (MA-PP) were used with the same glass fibers as reinforcement to produce UD composites. Fatigue damage in these composites was studied under tension-tension fatigue. The improvement in interfacial adhesion in MA-PP composite compared to PP composite was documented in a previous study [30]. Fatigue life in Region II showed an improvement of about a decade of fatigue life in the MA-PP composite. A closer examination of the fatigue damage mechanisms suggested that improved interface in MA-PP reduced the extent of debonding from broken fiber ends. This is expected to suppress the crack opening displacement of fiber-bridged matrix cracks and thereby reduce the crack growth rate.

The Region III of the FLD is below the fatigue limit where any fatigue damage if initiated evolves too slowly to lead to failure. The fatigue limit ε_{fl} depends on the fatigue limit of the matrix, as discussed above, and it is expected that it will be influenced by voids as enhancers of the local stresses. The fiber/matrix interface improvement discussed above will also tend to improve the fatigue limit.

7.4.1.2 Axial Compression–Compression and Tension–Compression Loading

The fatigue damage under axial cyclic compression was treated in Section 7.2.4. Accordingly, this case of damage is inherently dependent on fiber misalignment for its initiation as kink bands. Other manufacturing defects such as matrix voids and fiber/matrix interface imperfections can influence the growth of the kink bands. High-resolution XCT images, such as those shown in Chapter 5, display irregularities suggesting these effects. A detailed SEM study of kink band formation and growth in the presence of voids [31] also confirms the roles of voids in the initiation and growth phases of kink bands. The studies of compression failure of UD composites reporting the effects of voids have been done in static loading mode but the effects such as deflection of kink bands by voids and enhancing local shearing of the matrix to drive the kink bands can be expected for the cyclic loading case as well.

7.4.1.3 Transverse Tension–Tension Loading

In Chapter 5 we discussed the effect of matrix voids on the initiation of fiber/matrix debonding under transverse tension. In fatigue the debonds initiated in the first load application will grow in subsequent load cycles where the presence of the voids is expected to influence the growth process. At some point, the debonds kink out into the matrix and lead to the coalescence of debonds by linking with neighboring debonds. The interaction with voids in this process can be understood by a detailed stress and failure analysis which will be addressed in a later chapter. Experimental evidence suggests that the voids play a role in the early stages of transverse cracking by indicating the intersection of matrix cracks with voids such as the image in Figure 7.21 [32]. These images are for transverse cracks in 90° and −45° plies of glass/epoxy $[0/90_2]_s$ and $[0/45_2/0/−45_2]_s$ laminates, respectively.

7.4.1.4 In-Plane Shear and Combined Loading

Fatigue testing of UD composites under cyclic in-plane shear and under its combination with other loading modes is difficult to perform. Systematic studies of the effects of manufacturing defects on fatigue damage mechanisms are therefore not available. These damage mechanisms are instead interpreted by observing fatigue damage in laminates with off-axis plies. Such observations will be reviewed next.

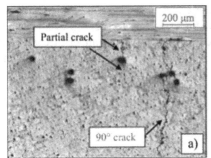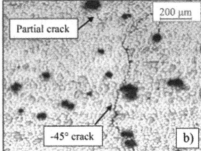

FIGURE 7.21 Optical micrographs showing evidence of voids connected to transverse cracks in $[0/90_2]_s$ and $[0/45_2/0/-45_2]_s$ glass/epoxy laminates. Source: Maragoni, L., Carraro, P.A., Peron, M. and Quaresimin, M., 2017. Fatigue behaviour of glass/epoxy laminates in the presence of voids. *International Journal of Fatigue*, 95, 18–28.

7.4.2 LAMINATES

7.4.2.1 Cross Ply Laminates

Cyclic tension loading of cross ply laminates in the axial direction allows studying fatigue damage in UD composites under transverse tension in the early stages of transverse crack formation and its multiplication under the constraint of the axial plies. This laminate configuration also allows for studying the initiation and growth of delamination caused by transverse cracks. These damage mechanisms have been described above in Section 7.3.1 and their relative roles have been assessed by the FLD framework. Here we shall focus on the observations related to the effects of manufacturing defects.

The manufacturing defects in cross ply laminates depend on the manufacturing methods that can be broadly described as autoclave curing of prepregs and resin infusion in a preform, e.g., RTM. The defects of interest in the context of fatigue damage in cross ply laminates are mainly voids in the matrix. The formation of voids was described in Chapter 2 and the methods for observing and characterizing them were described in Chapter 3. Here we shall review the studies that have investigated the role of voids in initiation of the first failure event in the 90° plies, the formation and evolution of the transverse cracks, and the subsequent delamination of the 0/90 plies. As noted above, the early stage of damage development is the same as in cyclic transverse tension of UD composites, but it is more conveniently studied by observation of the 90° plies within cross ply laminates. This is because a cross ply laminate avoids the abrupt failure of the UD composite by keeping it within the laminate. Further evolution of fatigue damage is then under the constraint of the 0° plies, as described above in Section 7.3.1.

The effect of voids on multiple transverse cracking in a cross ply laminate was studied in [32]. The authors manufactured $[0/90_2]_s$ laminates of glass/epoxy by VARTM and controlled the void content by using or not using degassing in the process. Two different void contents described by the area fractions of voids in the micrographs were used. To characterize the effect of voids on multiple transverse cracking a density measure used for partial cracks was used. Figure 7.22 shows how that crack density is affected by the presence of voids. The number of cycles that initiated a small number of cracks was also found to be affected by the void content.

Delamination caused by the transverse cracks on approaching the 0/90 ply interfaces was described above in Section 7.3.1 (Figure 7.15). Observing delamination in the interior of a laminate is not easy as most non-destructive methods observe external surfaces and miss this information. The delamination seen at the cut edges of a specimen can be misleading as it can be caused by the triaxial stress state existing at the free edges. By cutting the specimen longitudinally away from the free edge and observing the cut edge it is possible to verify if the delamination is what is depicted in Figure 7.15. Such an observation is shown in Figure 7.23 [33] that clarifies details that must be considered in modeling efforts. Note for instance the branching of the transverse crack and deflection into the 0/90 interface. The presence of local defects such as resin-rich areas, fiber distribution non-uniformity, and voids are possible causes for the unequal (unsymmetrical) size of the delamination on either side of the transverse crack. However, an experimental study to systematically delineate these effects does not appear to exist.

7.4.2.2 General Laminates

The FLD framework described above, if followed, can significantly reduce the burden of testing many possible laminate configurations. As described in Section 7.3.2, most characteristic features of fatigue damage become apparent by testing angle ply and quasi-isotropic laminates. These laminate configurations have also been used to clarify the effects of defects on fatigue damage. Some of the key observations will be reviewed next.

In the $[0/45_2/0/-45_2]_s$ glass/epoxy laminate tested under axial cyclic tension [32] the authors found that the formation of matrix cracks in the $-45°$ plies was influenced by the presence of voids (Figure 7.21), but the process of crack multiplication in these plies did not show the same sensitivity to voids as was the case for the transverse cracks in the $[0/90_2]_s$ laminate (Figure 7.22). They attributed this difference to the distribution of voids in the two laminate configurations possibly due to the processing procedures. Interestingly, the fatigue life for both laminate configurations was found to be affected, although less for the 45° case than for the 90° plies. It is noted that the delamination in both laminates was not studied and could hold an explanation for the differences found.

Depending on the processing method, e.g., RTM or autoclave curing, the distribution of voids will be different. While degassing in RTM will tend to remove voids from the entire volume of a panel, in autoclave curing the entrapped air between the prepregs will be the source of voids. Thus, the voids in the autoclave curing process will tend to reside in resin-rich layers and will induce stress concentration there on loading. Evidence of matrix cracks initiating from the 0/90 ply interfaces in a cross ply laminate processed by autoclave curing was reported in [34] and was described in Chapter 6. It has been argued that the observations made on specimen surfaces (faces and cut edges) do not necessarily represent the damage development in the

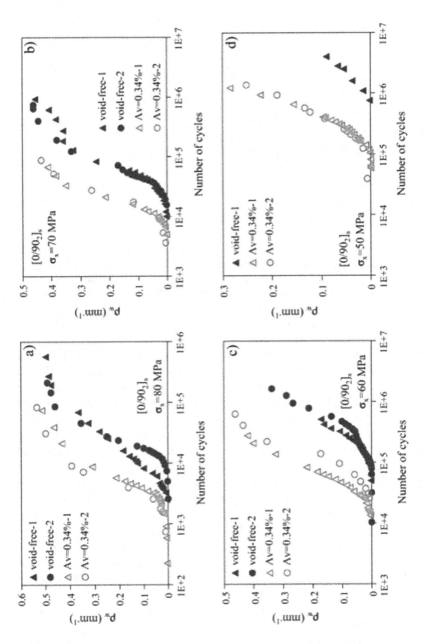

FIGURE 7.22 A weighted crack density for transverse cracks plotted against the number of load cycles in a glass/epoxy $[0/90_2]_s$ laminate at different applied load levels. Source: Maragoni, L., Carraro, P.A., Peron, M. and Quaresimin, M., 2017. Fatigue behaviour of glass/epoxy laminates in the presence of voids. *International Journal of Fatigue*, 95, 18–28.

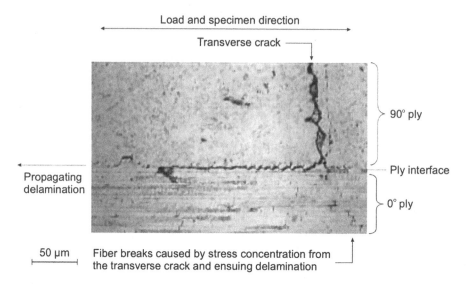

Load and specimen direction

Transverse crack

90° ply

Ply interface

Propagating
delamination

0° ply

50 μm   Fiber breaks caused by stress concentration from
             the transverse crack and ensuing delamination

FIGURE 7.23 A micrograph image showing the delamination at the transverse crack front
in a cross ply laminate that has been subjected a cyclic axial tension. Source: Gamstedt, E. K.,
Andersen, S. I., 2001. Fatigue degradation and failure of rotating composite structures-Materials characterisation and underlying mechanisms, Risø-R-1261, Risø National Laboratory,
Roskilde, Denmark.

interior of a composite laminate. While studies such as that reported in [22] were
able to show details of internal damage using X-ray radiography (Figure 7.15), the
effects of voids on damage events were not clarified. In the last decade, the use of
high-resolution CT has brought clarity to the role of voids in composite damage. One
noteworthy example of such studies is [35] where microfocus XCT was performed
to enable quantitative void analysis. Voids were quantified by their size, location,
and shape, and these quantitative measures were correlated with key features of
fatigue damage in an autoclave processed $[0/\pm45]_{3s}$ glass epoxy laminate subjected
to fully reversed axial loading. The void distribution was affected by the placement
of a resin-rich layer containing randomly oriented fibers, as indicated in the chart in
Figure 7.24 where the number of voids along the thickness of the laminate is plotted. This void distribution significantly biased the initiation and progression of the
fatigue damage. In any case, the methodology used in the study demonstrated the
capability of the XCT to provide details at a resolution needed to clarify the role of
voids and other defects of that scale. The study also clearly demonstrated the limitation of using straight specimens with cut edges to investigate fatigue damage where
the damage events within the volume (as in structures) are interfered with by the
free-edge-related damage.

   An important finding in [35] was that a stronger correlation exists between fatigue
damage in its early stage and the presence (location) of the largest void in comparison to the average void size. This should, however, be seen in the context of the voids

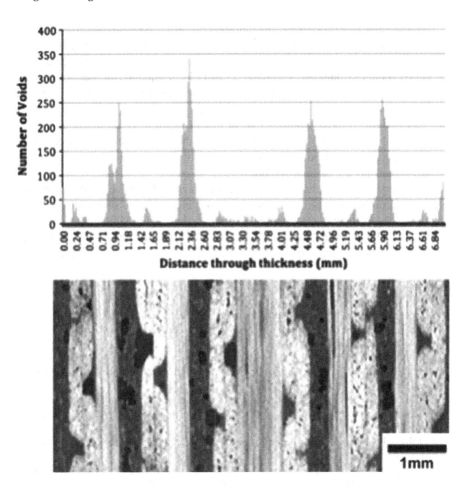

FIGURE 7.24   Chart showing the through-thickness void distribution, with a representative, through-thickness CT image aligned with the chart. Source: Lambert, J., Chambers, A.R., Sinclair, I. and Spearing, S.M., 2012. 3D damage characterisation and the role of voids in the fatigue of wind turbine blade materials. *Composites Science and Technology, 72*(2), 337–343.

in this study being mostly in the resin-rich layers placed intentionally rather than the resin-rich areas resulting from the processing, as discussed in Chapter 3. It is also worth noting that the findings of the study [35] should be seen keeping in mind the laminate configuration and the fully reversed loading. The compression-induced fatigue damage is likely affected by any fiber waviness which is inevitable in a practical manufacturing process.

7.5   CONCLUDING REMARKS

This chapter has focused on the effects of manufacturing defects on fatigue damage mechanisms. Before understanding these effects, it is necessary to have a clear picture of the fatigue damage process, i.e., its initiation, its progression, and its attainment of criticality leading to final failure. For this purpose, the fatigue life diagrams (FLDs) introduced by the author in 1981 [1] provide a suitable conceptual framework for the interpretation of damage. The baseline FLD for tension–tension fatigue of UD composites clarifies the non-progressive nature of the fiber-dominated failure at one extreme and the non-evolving damage providing the conditions for a fatigue limit at the other extreme. Proper modifications of this FLD allow for representing fatigue damage in other laminate configurations and for other loading modes. The effects of manufacturing defects are then better understood by focusing on the defects relevant to the mechanisms under consideration.

The observations of damage events have been reviewed in this chapter as reported in the literature noting the limitations and relevance of those observations. What cannot be observed and must be clarified by modeling and simulations has been pointed out. The following chapter, Chapter 9, will discuss the modeling and simulation efforts in view of the experimental observations.

REFERENCES

1. Talreja, R., 1981. Fatigue of composite materials: Damage mechanisms and fatigue-life diagrams. *Proceedings of the Royal Society of London. A. Mathematical and Physical Sciences*, 378(1775), pp. 461–475.
2. Talreja, R., 1982. *Damage Models for Fatigue of Composite Materials, Fatigue and Creep of Composite Materials*, Lilholt, H. and Talreja, R. (eds.), Risø National Laboratory, Roskilde, pp. 137–153.
3. Talreja, R., 1985. Conceptual framework for the interpretation of fatique damage mechanisms in composite materials. *Journal of Composites Technology & Research*, 7(1), pp. 25–29.
4. Talreja, R., 1987. *Fatigue of Composite Materials*. Technomic Publishing Co., Lancaster and Basel.
5. Talreja, R., 1989. Fatigue of composites. In Kelly, A., (ed.), *Concise Encyclopedia of Composite Materials*. Pergamon Press, Oxford, pp. 77–81.
6. Talreja, R., 1993. Fatigue of fiber composites. In Chou, T.W. (ed.), *Materials Science and Technology*. VCH Publishers, Weinheim, Chapter 13, pp. 584–607.
7. Talreja, R., 1995. A conceptual framework for interpretation of MMC fatigue. *Materials Science and Engineering: A*, 200(1–2), pp. 21–28.
8. Talreja, R., 2000. Fatigue of polymer matrix composites. In Talreja, R. and Månson, J-A. E. (Volume eds.), Kelly, A. and Zweben, C. (eds-in-Chief), *Comprehensive Composite Materials*, Vol. 2. Elsevier, Oxford, p. 529.
9. Talreja, R., 2003. Fatigue of composite materials. In Altenbach, H. and Becker, W. (eds.), *Modern Trends in Composite Laminates Mechanics*. Berlin: Springer, pp. 281–294.
10. Talreja, R., 2008. Damage and fatigue in composites–a personal account. *Composites Science and Technology*, 68(13), pp. 2585–2591.
11. Talreja, R. and Singh C.V., 2012. *Damage and Failure of Composite Materials*. Cambridge University Press, Cambridge.
12. Gamstedt, E.K. and Talreja, R., 1999. Fatigue damage mechanisms in unidirectional carbon-fibre-reinforced plastics. *Journal of Materials Science*, 34, pp. 2535–2546.
13. Castro, O., Carraro, P.A., Maragoni, L. and Quaresimin, M., 2019. Fatigue damage evolution in unidirectional glass/epoxy composites under a cyclic load. *Polymer Testing*, 74, pp. 216–224.

14. Garrett, K.W. and Bailey, J.E., 1977. Multiple transverse fracture in 90° cross-ply laminates of a glass fiber-reinforced polyester. *Journal of Materials Science, 12*, pp. 157–168.
15. Bailey, J.E., Curtis, P.T. and Parvizi, A., 1979. On the transverse cracking and longitudinal splitting behaviour of glass and carbon fibre reinforced epoxy cross ply laminates and the effect of Poisson and thermally generated strain. *Proceedings of the Royal Society of London. A. Mathematical and Physical Sciences, 366*(1727), pp. 599–623.
16. Harrison, R.P. and Bader, M.G., 1983. Damage development in CFRP laminates under monotonic and cyclic stressing. *Fibre Science and Technology, 18*(3), pp. 163–180.
17. Gamstedt, E.K. and Sjögren, B.A., 1999. Micromechanisms in tension-compression fatigue of composite laminates containing transverse plies. *Composites Science and Technology, 59*(2), pp. 167–178.
18. Quaresimin, M., Susmel, L. and Talreja, R., 2010. Fatigue behaviour and life assessment of composite laminates under multiaxial loadings. *International Journal of Fatigue, 32*(1), pp. 2–16.
19. Quaresimin, M. and Carraro, P.A., 2013. On the investigation of the biaxial fatigue behaviour of unidirectional composites. *Composites Part B: Engineering, 54*, pp. 200–208.
20. Quaresimin, M. and Carraro, P.A., 2014. Damage initiation and evolution in glass/epoxy tubes subjected to combined tension–torsion fatigue loading. *International Journal of Fatigue, 63*, pp. 25–35.
21. Plumtree, A. and Shi, L., 2002. Fatigue damage evolution in off-axis unidirectional CFRP. *International Journal of Fatigue, 24*(2–4), pp. 155–159.
22. Reifsnider, K.L. and Jamison, R., 1982. Fracture of fatigue-loaded composite laminates. *International Journal of Fatigue, 4*(4), pp. 187–197.
23. Grimes, G.C., 1977, January. Structural design significance of tension-tension fatigue data on composites. In *Composite Materials: Testing and Design (Fourth Conference)*. ASTM International.
24. Redon, O., 2000. Fatigue damage development and failure in unidirectional and angle-ply glass fibre/carbon fibre hybrid laminates. Technical Report Risø-R-1168, Risø National Laboratory, Roskilde.
25. Quaresimin, M., Carraro, P.A. and Maragoni, L., 2016. Early stage damage in off-axis plies under fatigue loading. *Composites Science and Technology, 128*, pp. 147–154.
26. Ryder, J.T. and Walker, E.K., 1977. Effect of compression on fatigue properties of quasi-isotropic graphite/epoxy composite. In K.L. Reifsnider and K.N. Lauratis (Eds.), *Fatigue of Filamentary Composite Materials*, STP 636, American Society for Testing and Materials, Philadelphia, pp. 3–26.
27. Quaresimin, M. and Carraro, P.A., 2013. On the investigation of the biaxial fatigue behaviour of unidirectional composites. *Composites Part B: Engineering, 54*, pp. 200–208.
28. Scott, A.E., Sinclair, I., Spearing, S.M., Mavrogordato, M.N. and Hepples, W., 2014. Influence of voids on damage mechanisms in carbon/epoxy composites determined via high resolution computed tomography. *Composites Science and Technology, 90*, pp. 147–153.
29. Gamstedt, E.K., Berglund, L.A. and Peijs, T., 1999. Fatigue mechanisms in unidirectional glass-fibre-reinforced polypropylene. *Composites Science and Technology, 59*(5), pp. 759–768.
30. Rijsdijk, H.A., Contant, M.A.A.J.M. and Peijs, A.A.J.M., 1993. Continuous-glass-fibre-reinforced polypropylene composites: I. Influence of maleic-anhydride-modified polypropylene on mechanical properties. *Composites Science and Technology, 48*(1–4), pp. 161–172.

31. Hapke, J., Gehrig, F., Huber, N., Schulte, K. and Lilleodden, E.T., 2011. Compressive failure of UD-CFRP containing void defects: In situ SEM microanalysis. *Composites Science and Technology*, *71*(9), pp. 1242–1249.
32. Maragoni, L., Carraro, P.A., Peron, M. and Quaresimin, M., 2017. Fatigue behaviour of glass/epoxy laminates in the presence of voids. *International Journal of Fatigue*, *95*, pp. 18–28.
33. Gamstedt, E. K. and Andersen, S. I., 2001. *Fatigue Degradation and Failure of Rotating Composite Structures-Materials Characterisation and Underlying Mechanisms, Risø-R-1261*, Risø National Laboratory, Roskilde.
34. Okabe, T., Imamura, H., Sato, Y., Higuchi, R., Koyanagi, J. and Talreja, R., 2015. Experimental and numerical studies of initial cracking in CFRP cross-ply laminates. *Composites Part A: Applied Science and Manufacturing*, *68*, pp. 81–89.
35. Lambert, J., Chambers, A.R., Sinclair, I. and Spearing, S.M., 2012. 3D damage characterisation and the role of voids in the fatigue of wind turbine blade materials. *Composites Science and Technology*, *72*(2), pp. 337–343.

8 Failure Modeling

8.1 INTRODUCTION

This chapter will deal with how to account for manufacturing-induced defects in composite materials toward assessments of their effects on the composite performance under static loading mode. The next chapter (Chapter 9) will address the modeling of failure under cyclic loading in the presence of defects. The previous chapters have described manufacturing methods (Chapter 2), quantification of manufacturing defects (Chapter 3), microstructural descriptors for simulating defects (Chapter 4), and observations of damage in the presence of manufacturing defects under static loading in UD composites (Chapter 5) and in laminates (Chapter 6). As described in the manufacturing sensitive design strategy in Chapter 1, the knowledge in Chapters 2–7 will form input to physical modeling for the purpose of assessing the structural performance.

Physical modeling (or physics-based, or mechanisms-based, modeling) of failure in composite materials must begin at the scale in the interior of the material where failure initiates. As described in Chapters 5–7, failure in composite materials consists of a sequence of events that culminate into a critical state of instability of failure progression or into an undesirable state from the performance perspective. An example of this is a laminate whose load-bearing capacity is lost abruptly (state of instability), or its deformational characteristics (stiffness coefficients) are degraded to an unacceptable level, for continued safe performance. To make such structural performance assessments, the modeling of failure events must follow failure progression to the prescribed criticality. In other words, the modeling approach should be able to increase the scale from the initial microstructural scale of failure initiation to the scale at which the criticality of the failure state occurs. This approach is generally described as "multi-scale modeling" and for composite materials it has been an active field in the past two decades if not more. However, there is no single version of this modeling approach and various computational simulation schemes have enriched this field into multiple paths for devising and connecting the scales. Our purpose here is to address multiscale modeling that is appropriate for the accounting of the manufacturing defects in the assessment of failure under different loading modes. It will be helpful, therefore, to first review the field of failure modeling to be clear on where it stands with regard to multiscale modeling. We shall begin with the early phenomenological failure theories and move to the recent developments in the physical modeling of failure. The author's previous reviews [1, 2] have also addressed this subject.

DOI: 10.1201/9781003225737-8

8.2 HOMOGENIZED COMPOSITE FAILURE THEORIES

Much before the failure events in composite materials were observed with sufficient clarity, attempts were made to describe the "strength" of UD composites by regarding these as homogeneous solids of anisotropic properties. In fact, the deformational response of a UD composite was commonly described in these works assuming linear stress–strain relations with directionally dependent elastic constants. Assumptions were then made that strength can similarly be described by certain directionally dependent material constants. The first notable effort to describe the failure of a homogenized UD composite appeared in 1965 by Azzi and Tsai [3], which has come to be known as the Tsai–Hill theory. In 1948, Hill [4] proposed a generalization of the yield criterion for isotropic metals to orthotropic metals by assuming six independent yield stresses, three for normal stresses in the three symmetry directions, and another three for shear stresses in the three planes of symmetry. It is noteworthy that while the von Mises criterion expressed in terms of stresses is the same as that using the distortional energy density criterion, this is not the case for the Hill criterion. Thus, Hill's generalization of the von Mises criterion in terms of stresses should be regarded as a mathematical generalization without resorting to the distortional energy density. However, this generalization can be justified on the assumption that the underlying mechanism of yielding does not change due to anisotropy. Hill in fact motivated his criterion from the need to describe yielding in metals that develop texture from sheet forming by rolling, resulting in the orthotropic symmetry, thereby rendering the yield stress directionally dependent without changing the underlying mechanism of yielding. The Azzi–Tsai theory used the same generalization of strength as Hill's generalization of metal yielding implying that there was a single failure mechanism in all loading modes. Our knowledge of the failure in UD composites (Chapter 5) clearly shows that the failure mechanisms are drastically different in different directions, e.g., in axial tension and transverse tension. Thus, the Tsai–Hill failure criterion for UD composites unlike the Hill criterion for yielding in orthotropic metals does not have a basis in the physical mechanisms.

The Tsai–Hill failure criterion for in-plane stresses after assuming transverse isotropy in the cross-sectional plane of a UD composite takes the following form.

$$\frac{\sigma_{xx}^2}{X^2} + \frac{\sigma_{yy}^2}{Y^2} - \frac{\sigma_{xx}\sigma_{yy}}{X^2} + \frac{\tau_{xy}^2}{T^2} = 1 \tag{8.1}$$

where σ_{xx} and σ_{yy} are the normal stresses in the fiber and transverse directions, respectively, τ_{xy} is the in-plane shear stress, and X, Y, and S are the strengths, i.e., the maximum attained values of σ_{xx}, σ_{yy}, and τ_{xy}, respectively. For unequal strengths in tension and compression, a variation of Equation (8.1) is easily derived.

The next noteworthy development of failure theories for UD composites appeared in a paper by Tsai and Wu [5] in 1971 based on a formulation proposed by Gol'denblat and Kopnov [6] in 1965. The proposed formulation (criterion) is in terms of a scalar function expressed as a polynomial of all stress tensor components. The coefficients of the polynomial terms represent the strength constants. The Tsai–Wu criterion is a simplification of the Gol'denblat–Kopnov function to a quadratic expression that allows a graphical interpretation as a quadric surface in the stress coordinates. Thus, the ellipsoidal surface of "failure" of a UD composite under in-plane stresses takes the form,

$$F_p \sigma_p + F_{pq} \sigma_p \sigma_q = 1 \qquad (8.2)$$

where the indices p and q take values 1, 2, and 6, representing the indices xx, yy, and xy, respectively, in the usual notation, and the summation convention for repeated indices is implied. The coefficient terms in the equation represent the inverse of the strength values of the corresponding stresses, as in the Tsai–Hill criterion, Equation (8.1). The coefficients of the linear terms are non-zero when the strengths for positive and negative stress components differ. Thus, F_6 is always zero, and for the same reason, F_{16} and F_{26} vanish. Of the remaining coefficients, F_1, F_2, F_{11}, F_{22}, and F_{66} can be expressed in terms of the normal strength values in the fiber and transverse directions and the shear strength in the plane of the composite. Finally, the coefficient F_{12} is an "interactive" term that in principle can be obtained, e.g., by a biaxial test. However, the value obtained for F_{12} depends on the method used, and the non-uniqueness thus induced is a source of ambiguity in the criterion. It can be argued that since the Tsai–Wu criterion, Equation (8.2), cannot be related to the specifics of the failure mechanisms in composites, its applicability will always be subject to uncertainty, additionally worsened by the non-unique value of the F_{12} coefficient.

At this point, it can be said that the two criteria discussed above are severely handicapped by their inherent deficiencies, namely, the Tsai–Hill criterion is motivated by an incorrect failure mechanism (yielding, which is not the failure mechanism in UD composites with a non-metallic matrix), and the Tsai–Wu criterion is simply a quadratic curve-fitting framework having no basis in the failure mechanisms (initiation, progression, and criticality).

Further developments in formulating failure theories for UD composites took place when Hashin [7] in 1980 pointed out that the single ellipsoid represented by Equation (8.2) led to physically unacceptable interactions between stress components in some cases. He suggested introducing piecewise smooth surfaces to describe critical failure states for mitigating those anomalies. Additionally, Hashin [7] suggested recognizing that composite failure involving fiber breakage was governed by stresses differently than when the failure occurs in the matrix only. Assuming the two failure modes to be independent, Hashin [7] proceeded to formulate criteria for them separately. Thus, for example, the fiber failure mode when the fiber axial stress is tensile was assumed to be given by,

$$\left(\frac{\sigma_1}{X}\right)^2 + \left(\frac{\sigma_6}{T}\right)^2 = 1 \qquad (8.3)$$

Thus, the transverse normal stress was assumed not to affect the fiber failure mode and the quadratic form of the failure criterion was retained.

For the matrix failure mode, Hashin [7] proposed the notion of a failure plane not intersecting fibers in a UD composite. Assuming the fiber direction stress not to have an influence on this plane, he proposed that the other stress components would determine the inclination of the plane. However, he noted that there was insufficient knowledge of failure for determining the angle of this inclination and proposed that some (yet unknown) extremum principle should govern it. Notwithstanding Hashin's hesitation to proceed with the idea of a matrix failure plane, Puck and his associates [8, 9] launched an elaborate procedure to develop failure criteria incorporating the failure plane idea. The resulting failure criteria for a UD composite require the determination of at least seven material constants, not all of which can be interpreted as material properties.

The notion of a failure plane forming in the matrix of a UD composite, as proposed by Hashin [7], focuses on the final failure event, ignoring the initiation and progression of failure preceding it. Failure always initiates at a point and must therefore by addressed by a *stress state* whose criticality causes that failure. Formulating a failure criterion in terms of the traction vector on a plane cannot account for the initiation of failure unless the traction vector acts *uniformly* on the plane, which cannot occur generally for a finite-sized plane. Furthermore, while one can argue that initiation of failure at a point in the matrix of a UD composite will subsequently result in the formation of a failure plane, this argument will hold only for an unconstrained UD composite. When a UD composite acts as a ply within a laminate, failure in the matrix is constrained by the neighboring plies, and depending on the severity of the constraint, the initiation of failure in the matrix will lead to a sequence of failure events before the formation of the failure plane. The details of this failure process cannot be treated if the UD composite is homogenized as is the case in the Hashin failure theory.

Hashin [7] noted the limitations in his failure theory by stating that his proposed "mathematical modeling of the failure modes…should not be considered definitive in any sense". He admitted that the failure modeling was "merely a simple mathematical description" based on the experimental evidence. At the time of Hashin's publication of the failure theory (1980) the experimental evidence was not sufficient to perform detailed analyses of the failure modes in matrix and fibers. The separation of failure modes in matrix and fiber failures, however, avoided the need to determine the interaction coefficient F_{12} appearing in the Tsai–Wu criterion. In Christensen [10], a theory put forth later, the F_{12} coefficient is also avoided by separating the fiber and matrix failure modes. In this theory, Christensen used the stress invariants for transversely isotropic UD composites and derived a single expression for the matrix failure mode against two separate failure criteria for tension and compression matrix failure modes in Hashin [7]. The final form of the Christensen criterion can be seen as a reduced form of the Tsai–Wu criterion, as noted in [11].

There have been numerous failure criteria proposed following Hashin's failure mode separation concept and the hypothesis of a failure plane in the matrix [7]. Other than the Puck criteria mentioned above, the failure plane was further explored by making phenomenological assumptions resulting in variations and modifications of the Puck criteria denoted as the LaRC criteria, labeling the criteria by the affiliation of researchers (NASA Langley Research Center). The latest version of these criteria is the fifth version, LaRC05 [12].

When assessing the validity of any proposed failure theory, it has been commonly implied that the description given by the theory should agree with experimental data. Since such data usually record only the occurrence of the final failure event and denote the associated imposed uniform stress as "strength", the validation of a homogenized failure theory is only with respect to the final event. Each proposed failure theory uses the strength values obtained under a single component of stress, e.g., under a uniaxial normal stress or under an in-plane shear stress, and predicts failure under a combination of those stresses. The predictions are presented in the form of "failure envelopes" that connect the combined stresses at failure by continuous curves in one or more quadrants in such plots. Experimental data generated under combined loading are then placed alongside the predictions on the failure envelope plots for comparison. The uncertainties involved in testing and the randomness of the failure process produce scatter in the test data, and it is, therefore, common to plot averages of strength data if multiple tests have been conducted. It is important to note that the predictions of the failure theories are often deterministic, while the recorded test data involve randomness. Many comparisons have been reported over the years, a common feature being that none of the theories agree with test data in all cases. Figure 8.1 is a recent example where four commonly cited homogenized failure theories are compared with experimental data [11].

As seen in Figure 8.1, none of the theories agree with the test data to the same degree for all four materials. As noted above, comparing the random test data against

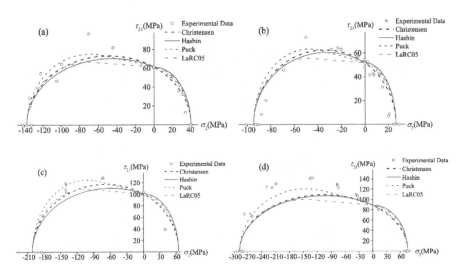

FIGURE 8.1 Failure envelopes described by four failure theories in combined transverse normal stress (s_2) and shear stress (t_{12}) compared with data for (a) E-glass/LY556, (b) AS4/55A, (c) T800/3900-2, and (d) IM7/8552. Source: Gu, J., Chen, P., Su, L. and Li, K., 2021. A theoretical and experimental assessment of 3D macroscopic failure criteria for predicting pure inter-fiber fracture of transversely isotropic UD composites. *Composite Structures*, 259, 113466.

the deterministic predictions of the theories makes it difficult to infer the validity of the theories. However, the deviation of the predictions from test data for combinations of high shear stress and low transverse compressive stress is particularly striking. The apparent shear "strengthening" by transverse compression can be made to agree with predictions to some extent by adjusting the inclination of the failure plane in the Puck and LaRC criteria. Figure 8.2 illustrates the angle adjustment procedure to fit a given set of test data [12].

The data in Figure 8.1 suggest that a single smooth curve is not able to fit the test values, resulting in some of the data becoming "outliers". This questions the underlying assumption that a single failure mechanism exists and that the failure results in a failure plane whose inclination depends on the combined effect of normal and shear tractions on that plane. It seems more appropriate to assume the presence of multiple failure mechanisms, each reaching its criticality under the combined effect of the local stresses in the composite. Under this perspective, the failure envelope in the (τ_{12},−σ_2) quadrant in Figure 8.1 will take the form depicted in Figure 8.3, where the two failure mechanisms are labeled as mechanism 1 and mechanism 2, corresponding to the transverse compression failure and shear failure, respectively. The single mechanism conventionally assumed in homogenized failure theories is also shown in Figure 8.3. The failure envelope parts for the two mechanisms are drawn

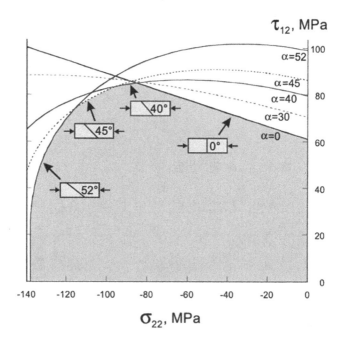

FIGURE 8.2 The inclination of the failure plane in a UD composite changes the predicted failure under combined transverse normal compression and in-plane shear stress. Source: Davila, C.G., Camanho, P., 2003. Failure criteria for FRP laminates in plane stress. NASA/TM-2003-212663.

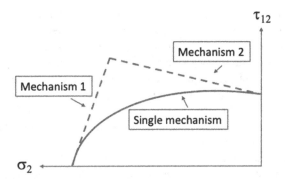

FIGURE 8.3 A possible scenario of two failure mechanisms under combined shear and transverse compression imposed on a UD composite. Failure theories describe the failure interaction with a continuous curve implying a single mechanism.

as straight lines only for illustration. Obviously, they will be nonlinear in a general case. It is to be noted, however, that the prediction of the two parts in the failure envelope will not be a simple matter in the homogenized failure theories. This is because in such theories the *local* stress states responsible for the criticality of the failure mechanisms do not enter. To appropriately address the failure mechanisms, therefore, one must resort to failure analyses that are conducted at the scales at which failure events occur. This is done in multiscale failure modeling which will be discussed next.

8.3   MECHANISMS-BASED MULTISCALE FAILURE MODELING

Heterogeneous solids such as composite materials lend themselves to analyses at multiple scales. The need for a multiscale analysis emerged initially for estimating overall (global, macroscale) properties from the knowledge of the properties of the constituents and their arrangement at the microstructural scale. A classical paper by Hill [13] in 1963 laid the theoretical foundation for estimating the overall elastic moduli of a mixture of two firmly bonded phases. In that paper, Hill introduced the concept of a representative volume element (RVE) that has become integral to all multiscale analyses.

In Chapter 3 we treated the quantification of manufacturing defects at a microstructural scale and in that context, we discussed the need to find a representative volume (or region) over which statistical averages of defect characteristics can be formulated. The RVE concept was explored further for nonuniform distributions of microstructural entities in Chapter 4. There, the scale of disorder was discussed as a measure of the extent of the region needed for the representation of the nonuniformity of the microstructural entities. The statistical descriptors thus formulated could be used to generate simulations (called realizations, i.e., representative samples) of the microstructure. While measures such as the scale of disorder allow the geometrical representation of the microstructure, we need to be sure that RVE realizations will generate representative local stress fields when the RVE bounding surfaces are subjected to a prescribed loading. To address this issue, it seems reasonable to resort to Hill's RVE concept [13] which requires the "apparent overall moduli to be effectively independent of the surface values of tractions and displacements". While this intuitive requirement of the minimum RVE size is for elastic properties, it seems to be a good guideline for assuring proper transfer of external impulses to local regions of a heterogeneous solid. Therefore, this approach will be pursued here for generating local stress fields within the RVE of a composite for conducting failure analysis.

The multiscale modeling field for composite materials has been active for some time. Several reviews and collections of articles [14–16] are available. Our focus here is on incorporating manufacturing defects in multiscale failure analyses. In the following, therefore, failure mechanisms in the presence of manufacturing defects as observed and described in Chapter 5 for UD composites and in Chapter 6 for laminates will be considered in relation to appropriate multiscale modeling and simulation approaches. Before considering defects, however, the generic features of failure modeling will be reviewed.

8.4 FAILURE IN UD COMPOSITES

8.4.1 LONGITUDINAL TENSION

8.4.1.1 Fiber Breakage and Fiber–Matrix Debonding

As described in Chapter 5, Section 5.2, fiber failure is a statistical process because of manufacturing defects such as nonuniform fiber diameter and random surface flaws created by surface treatments (sizing). When the fibers are embedded in a polymer matrix, additional flaws occur such as broken fibers during handling of fibers and incomplete wetting of fiber surfaces that can produce dry regions leading to imperfect fiber–matrix bonding. Additionally, voids can be present in the matrix produced by volatiles during curing and if any entrapped air is not removed.

On loading a UD composite in tension along fibers, two scenarios are possible as a first failure event: 1) Preexisting broken fibers cause stress concentration in neighboring fibers leading to new fiber breaks and 2) fibers break at random locations when the local tensile stress at these locations exceeds the failure stress. In most studies, only the second scenario is considered and models for prediction of tensile strength are developed. A recent review assessed a few such strength models [17] summarizing a common understanding of the failure process beyond the first failure event as follows. "The load of a broken fibre is shed to the nearby fibres, which are hence subjected to stress concentrations; this increases their failure probability and causes a tendency to develop clusters of fibre-breaks, which increases stress concentrations even further. At some point, one cluster will become so large that it starts growing unstably; this critical cluster will cause final failure of the composite".

The key points in the failure process described above can be summarized as: 1) Statistical nature of fiber failure and 2) formation and enlargement of the fiber failure clusters. Most models in the literature have focused on the statistical nature of fiber failure but have not fully addressed the fiber failure cluster formation.

Specifically, the roles of the matrix and the fiber–matrix interface in causing additional fiber failures in the presence of the first fiber failure have not been accounted for to the full extent. The early treatments of the load transfer from a broken fiber to a neighboring intact fiber used the shear-lag approach assuming a perfectly bonded fiber–matrix interface and an elastic matrix [18, 19]. An extension to the elastoplastic matrix was proposed in [20] and in [21] the interfacial slip was considered. The fact that the critical fiber breakage cluster must grow unstably to cause final failure suggests that fibers within that cluster must be interconnected by a matrix crack with accompanying fiber–matrix debonding. The fiber–matrix debonding is necessary to allow connecting broken fiber ends in a single plane that will grow unstably to cause total failure. Such an uneven fracture plane formed by a matrix crack was assumed as the background for the analysis conducted in [22]. That analysis will be described below highlighting the results and their implications, leaving full details to [22]. Before doing that, however, one misconception regarding the role of matrix cracks in the failure process resulting from another study [23] needs to be addressed. In that study, the presence of matrix cracks on stress concentrations on fibers was found to be negligible. It should be noted that this inference resulted from the assumed placement of a matrix crack in the inter-fiber region. In that situation, the matrix crack could not conduct any significant surface opening under loading due to the heavy constraint by the surrounding stiff fibers. It is unlikely that a matrix crack forms under axial tension without fibers playing a role. One realistic scenario is where a fiber breaks under tension causing fiber–matrix debonding at the broken end which then leads to the formation of a matrix crack by the debond crack kinking out of the interface. In a thorough study [24] that performed fiber fragmentation tests on single-fiber glass–epoxy and carbon–epoxy composites the nature of fiber–matrix debonding was clarified and the difficulty of measuring the debond length was discussed. Figure 8.4 taken from that study illustrates how the fiber break gap and the debond zone are different in the loaded and unloaded states, emphasizing the importance of measuring the debond length under loading. The debond length can indeed be small for well-bonded composites and can be missed if not observed in the loaded state.

Furthermore, as described in Chapter 5, the extent of debonding from a broken fiber end depends on the quality of adhesion between fibers and matrix. For a well-bonded interface the observations in [25] indicate short debond length leading to the formation of a matrix crack. It should be noted that the resolution of the observations is often not enough to discern whether the matrix crack is a result of the debond crack kinking out in the matrix or whether the matrix crack forms without debonding. Such questions can be resolved by proper failure analysis. This was done in [22] and the results of that study follow.

The study in [22] considered two scenarios for stress concentrations on nearby fibers resulting either from a pre-existing broken fiber (Scenario 1) or from a fiber broken under loading (Scenario 2). The geometrical model following previous studies [26, 27] is depicted in Figure 8.5. As seen, a local region of the UD composite cross section is considered a nucleation site for the fracture plane. This region consists of a broken fiber at the center of six hexagonally arranged fibers. To calculate local

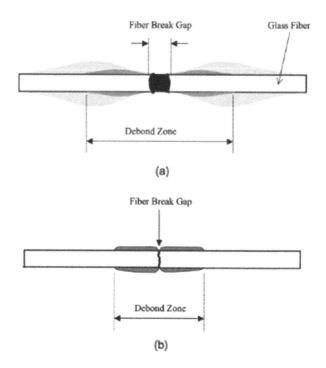

FIGURE 8.4 Schematic illustration of the fiber break and the associated debond zone as
seen by photoelastic birefringence in (a) loaded state and (b) unloaded state. Source: Kim,
B.W. and Nairn, J.A., 2002. Observations of fiber fracture and interfacial debonding phenomena using the fragmentation test in single fiber composites. *Journal of Composite Materials*,
36(15), 1825–1858.

stresses induced by the broken fiber, the intact fibers are smeared together in a ring.
The surrounding composite is also modeled as a homogeneous region (Figure 8.5).
A finite element model is then used to compute stresses in the matrix region between
the central broken fiber and the homogeneous ring of neighboring fibers accounting
for the thermal cooldown-induced residual stresses and stress singularity effect at
the broken fiber end [22].

Scenario 1: Pre-existing broken fiber
In this scenario, the stress field calculated in the epoxy matrix next to the pre-
existing broken fiber shows triaxiality and a high level of dilatational energy density
in a close vicinity of the broken fiber end. According to the studies by Asp et al
[28, 29], this leads to cavitation-induced brittle cracking when the dilatation energy
density exceeds the experimentally measured critical value for epoxy. The distortional energy density at that load level was found not to reach the value needed to
induce yielding. Thus, the first failure event is concluded to be a short brittle crack
induced by a cavity formed in the matrix. It is noted that cavitation in glassy polymers is a highly local phenomenon that requires equal (or nearly equal) principal

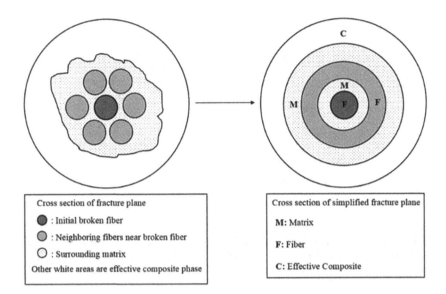

FIGURE 8.5 An axisymmetrical model for analyzing failure from a broken fiber. Source: Zhuang, L., Talreja, R. and Varna, J., 2016. Tensile failure of unidirectional composites from a local fracture plane. *Composites Science and Technology*, *133*, 119–127.

stresses [30, 31]. Therefore, the initial crack formed by the cavitation is expected to be small. The computed stress field also shows that the matrix in the inter-fiber region has significant triaxiality suggesting a low level of inelastic deformation. The crack formed by cavitation is therefore expected to grow in an approximately elastic field. The brittle cavitation-induced debonding will be discussed in more detail later in Section 8.4.3.

Assuming a short crack near the broken fiber end, its further growth was calculated by the extended finite element method (XFEM) [22]. The crack path is displayed in Figure 8.6. As seen, the crack grows transverse to the fiber axis. This crack on meeting a neighboring fiber will likely debond the fiber. However, in the model geometry employed in the study, this failure process cannot be analyzed because of the way the neighboring fibers are represented, i.e., as a homogeneous cylinder representing all six fibers. An average stress enhancement in the neighboring fibers can, however, be estimated following previous procedures used for the same model geometry [26]. That stress enhancement is then used to calculate the increased probability of failure of the intact neighboring fibers. The estimates of the enhanced probability of failure depend on the fiber length over which the increased stress is taken. For a short length equal to two fiber diameters, the relative increase in the probability of failure came out to be by a factor of 1.67. While the actual increase in the failure probability may not be accurate because of the homogenization of the intact fibers, the fact that the probability of fiber failure is high for short fiber lengths seems to be corroborated by the experimental findings of an XCT reported in [32].

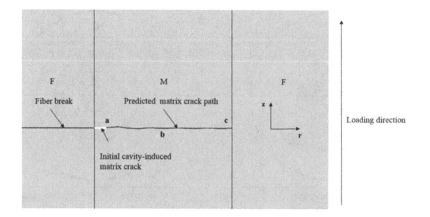

FIGURE 8.6   The path of an initial crack formed by cavitation in the matrix near the pre-existing broken fiber end. Source: Zhuang, L., Talreja, R. and Varna, J., 2016. Tensile failure of unidirectional composites from a local fracture plane. *Composites Science and Technology*, *133*, 119–127.

Scenario 2: Fiber break during loading

As the experimental observations have shown (see Chapter 5), a UD composite in longitudinal tension shows distributed single-fiber breaks called singlets at low load levels. Some of the singlets on further loading induce neighboring fiber breaks described as doublets. The fiber failure process shows clusters of fiber breaks that have led to the notion of a critical fiber cluster that grows unstably to failure, as described above. The details of the fiber cluster growth are not easy to observe even with a high-resolution XCT. Simulation studies are therefore necessary to reveal a plausible process of fiber cluster formation and growth. In the approach taken in [22], it is recognized that when a fiber breaks, it must debond the fiber. The length of the debond crack along the fiber is expected to depend on the fiber–matrix interfacial characteristics. Various models have attributed certain characteristics (strength, fracture toughness, etc.) to an interface that depends on the modeling assumptions. For instance, if the interface is assumed as a surface (of zero thickness), then the debonding is taken as an interface crack whose resistance to growth is defined as "fracture toughness" (without a material), and the well-established fracture mechanics methodology for such cracks is used. Alternatively, cohesive zones are assumed for which traction–separation relations are employed with or without incorporating friction. A recent effort [33] has studied the effects of assumed interfacial characteristics on the debonding where citations of previous models can be found. How the debonding of the fiber–matrix interface is assessed depends on what properties are assumed for the interface. The uncertainty in this approach is obvious when one realizes that the properties are seldom found by independent experiments and are often derived (back calculated or calibrated) from the test results.

A somewhat unconventional view of the fiber–matrix interface was taken in [22] that applies to relatively strong (well-bonded) interfaces such as those in glass–epoxy and carbon–epoxy composites. Instead of placing a debond crack at the interface, it is placed at a short distance away from the fiber surface, but parallel to it, in the matrix. This way the uncertainty of the interface characteristics (e.g., fracture toughness of zero-thickness material) is avoided; instead, the better-defined properties of the matrix material are used. The fact that failure occurs in the matrix close to the interface in composites with well-bonded interfaces is supported by the observed matrix residue found on the fiber surfaces. However, because of the short debond lengths in well-bonded interfaces, it is difficult to find clear evidence of failure at the interface versus in the matrix. An indirect but clear evidence is seen in Figure 8.7 where images of pulled-out fibers after the completion of microbond tests are shown [34]. The clean surface seen for fiber surfaces marked "category A" indicates interfacial failure, while "category B" refers to failure where a residual resin is seen on the fiber surface indicating failure in the matrix. The authors attributed the matrix failure to the degradation of the polymer induced by curing. In any case, it can be inferred that if the interface is well-bonded, then the failure is likely to occur in the thin layer of the matrix that has a lower resistance to failure than that provided by the interface. For modeling purposes, assuming failure in a thin resin layer has the advantage that uncertain properties of an ill-defined interface need not be determined.

Taking the approach of interface failure modeled as a short crack in the matrix of concentric cylindrical shape, as sketched in Figure 8.8, showing the longitudinal section used in the FEM model, the kink out of the crack into the inter-fiber region was analyzed. Based on the principle that the debond crack will kink out at an angle that will give the most driving force (energy release rate), the kink-out angle was calculated. This angle turned out to be negligibly sensitive to the debond length, and its value averaged over a range of short debonds was found to be 78.65° measured from the fiber axis. As an assurance that the modeling of debonding as a longitudinal

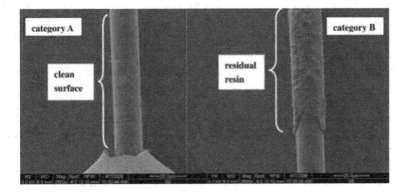

FIGURE 8.7   Failure after fiber–matrix debonding: Category A shows a clean fiber surface, and category B shows a residual resin on the fiber surface. Source: Yang, L. and Thomason, J.L., 2010. Interface strength in glass fibre–polypropylene measured using the fibre pull-out and microbond methods. *Composites Part A: Applied Science and Manufacturing*, 41(9), 1077–1083.

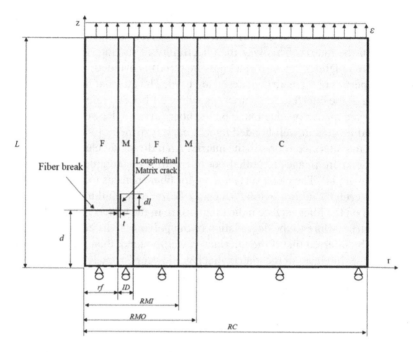

FIGURE 8.8 Longitudinal section of the axisymmetrical FEM model showing a longitudinal matrix crack placed concentrically with the central fiber at a distance $t = 0.1$ mm from the fiber surface. Source: Zhuang, L., Talreja, R. and Varna, J., 2016. Tensile failure of unidirectional composites from a local fracture plane. *Composites Science and Technology, 133*, 119–127.

matrix was not inaccurate, the kink-out angles calculated by using a well-established fracture mechanics approach for cracks at bi-material interfaces [35] were found to be in good agreement [22].

Figure 8.9 shows the kink out of the debond cracks into the matrix for different debond lengths and their further growth toward the neighboring fibers. Note that the neighboring fibers in this model are homogenized into a concentric cylinder and, as in Scenario 1, this does not allow analyzing the debonding of those fibers. The angle at which the cracks approach the neighboring fibers suggests the level of stress enhancement at the fiber–matrix interface. Thus, as indicated in Figure 8.9, shorter kinked-out debond cracks will induce more debonding in the neighboring fibers than will be the case for longer debonds.

The model in [22] goes as far as analyzing the initial stage of formation of the failure plane. As noted above, the debonding and failure of the neighboring fibers cannot be addressed by the model because of the homogenization done by those fibers. This needs to be done to create interconnectivity of the broken fibers by a plane that will eventually grow unstably to cause the UD composite failure. The models in the literature have ignored such a plane and have instead focused on fiber break density [36] to find the critical cluster size for final failure. The pursuit of this direction of

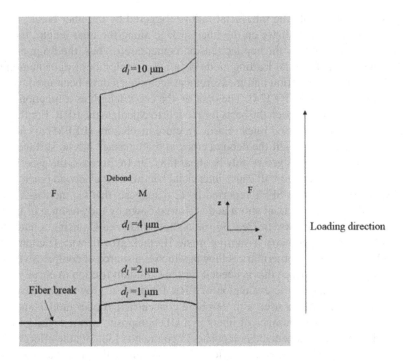

FIGURE 8.9   Kink out of debonds cracks of different lengths (d_l) into the matrix. Source: Zhuang, L., Talreja, R. and Varna, J., 2016. Tensile failure of unidirectional composites from a local fracture plane. *Composites Science and Technology*, *133*, 119–127.

research has been influenced by the early work on local load sharing and the associated failure probabilities of the load-sharing fibers [37]. Those early works used simple concepts such as shear-lag for stress transfer from a broken fiber to a neighboring intact fiber. Proper stress analysis in the vicinity of a broken fiber shows stress triaxiality in the matrix followed by fiber/matrix debonding, kink out of the debond crack, and a nonlinear crack path, as reported in [22]. However, the stress concentration factor, as commonly defined and used in [22], is an inadequate measure of the effect of a broken fiber on the surrounding fibers for the purpose of analyzing the formation of a broken fiber cluster. Clearly, a cluster of discrete fiber failures cannot reach instability unless interconnected by a matrix failure plane. More work is therefore needed to reexamine the formation and growth of broken fiber clusters.

8.4.1.2   Growth of Fiber–Matrix Debonding

Once a fiber–matrix debond crack has initiated from a broken fiber end under axial tension, its further growth along the fiber on increased loading is of interest and has been a subject of study by many researchers. Based on fracture mechanics principles, the objective has been to determine the energy release rate (ERR) as a driving force for the debond crack growth. Most early approaches to this problem used a single fiber embedded in a polymer matrix and applied axial tension to this composite to investigate the interfacial failure triggered by the fiber break. Due to the presence of multiple flaws on the fiber surface along the fiber length, the fiber breaks at multiple sites as the applied tension is increased. Thus, the fragmentation of fiber in this experimental loading mode is called a single-fiber fragmentation test (SFFT). Various observation and measurement techniques have been used to track the debond growth in SFFT [24]. The data on the crack length as a function of the applied tension has then been the basis for models to calculate the ERR. For this purpose, one approach has used linear elastic fracture mechanics (LEFM) to calculate the mode II ERR, i.e., G_{II}, if the debond crack at the interface has no surface opening displacement and thus grows only be shear [38, 39]. In this case, the asperities of the new surfaces generated will cause interfacial friction as the debond crack grows. The experimental data in SFFT reported by [24] suggests that G_{II} increases nearly linearly with the applied axial strain and is slowed down by the presence of friction as the debond crack grows. In [38, 39] the authors analyzed G_{II} using a numerical model of the SFFT geometry assuming mode II crack growth with friction. They included in their analysis thermal cooldown, which is a source of compressive stress at the fiber/matrix interface that is needed for the Coulomb friction to occur.

Another approach to the evaluation of ERR for debond growth was developed in [40, 41]. These authors used a three-phase concentric cylinder model, shown in Figure 8.10, where the presence of fibers in a UD composite was recognized in the outer cylinder as a homogenized composite. In contrast to [38, 39], the effect of friction on the debonded surfaces was neglected. The G_{II} thus calculated was lower than what was obtained in a single-fiber model [39, 40] due to the constraining effect of the surrounding composite that was accounted for in the model.

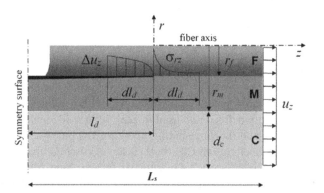

FIGURE 8.10 A three-phase axisymmetrical model of length L_s used in [40, 41]. The fiber (F) is debonded from the matrix (M) over a length l_d. The relative displacement Δu_z of the fiber in the debonded region and the interfacial shear stress σ_{rz} ahead of the debond crack tip over an incremental debond length dl_d are indicated schematically. The radial distances r_f and r_m to the boundary of the fiber and the matrix, respectively, and the diameter d_c of the homogenized composite (C) are indicated. Source: Zhuang, L., Pupurs, A., Varna, J. and Ayadi, Z., 2016. Effect of fiber clustering on debond growth energy release rate in UD composites with hexagonal packing. *Engineering Fracture Mechanics, 161*, 76–88.

The calculated G_{II} showed enhancement over a short distance from the broken fiber end due to the perturbation in the crack tip stress field caused by the energy released in breaking the fiber. Beyond this distance, a steady-state ERR was found. It should be noted that this result does not account for any friction between the debonded surfaces, which would tend to increase the G_{II} value with the debond crack length.

In real UD composites, the fiber distribution in the cross section is nonuniform. To account for the fiber clustering effect on the debond crack ERR, an axisymmetric model with a core containing a broken and partially debonded fiber at the center with six discrete surrounding fibers placed in a hexagonal pattern was used in [42]. The outer cylinder of the model was a homogenized composite, as depicted in Figure 8.11.

A detailed stress analysis of the unit cell of the model in Figure 8.11 showed that at high fiber clustering, i.e., as the intact fibers got closer to the central broken fiber, the debond crack surfaces opened, suggesting a lack of friction. This opening had a distribution in the circumferential direction of the debonded fiber, which was found to be different under axial tension than under thermal cooling. However, the thermal effect was found to be much smaller than the mechanical effect.

8.4.2 LONGITUDINAL COMPRESSION

The failure mechanisms observed under longitudinal compression of UD composites were reviewed in Chapter 5, Section 5.3. As described there, the main mechanism governing failure observed is fiber microbuckling leading to kink-band formation. The role of fiber misalignment (or fiber waviness) was clarified by observations at different scales of observations, ranging from SEM to XCT. The kink-band geometry based on the observations is repeated here in Figure 8.12. The discussion below will focus on modeling considerations of compression failure.

The earliest model for compressive failure of UD composites was proposed by Rosen [43] in 1965, which assumed two microbuckling modes in initially straight fibers, as depicted in Figure 8.12. The out-of-phase mode, called extensional mode,

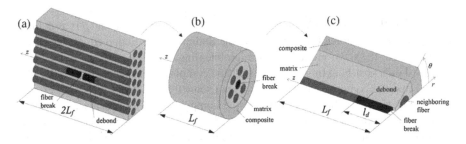

FIGURE 8.11 (a) A UD composite with one broken and debonded fiber, (b) a model representing the surrounding of the broken fiber, and (c) a unit cell of $\theta = 30°$ for finite element analysis. Source: Zhuang, L., Pupurs, A., Varna, J. and Ayadi, Z., 2016. Effect of fiber clustering on debond growth energy release rate in UD composites with hexagonal packing. *Engineering Fracture Mechanics*, *161*, 76–88.

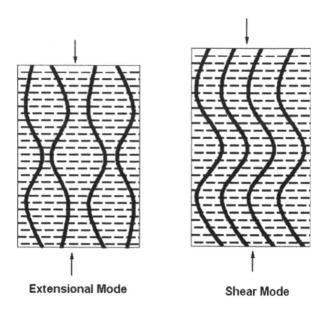

Extensional Mode   **Shear Mode**

FIGURE 8.12 Extensional (out-of-phase) and shear (in-phase) microbuckling modes as postulated by Rosen.

gave higher compressive strength than the in-phase mode, called shear mode. The prediction of compressive strength by the shear mode of fiber microbuckling, $\sigma_c = G$, the shear modulus of the UD composite, was found to over-estimate the experimentally found values.

The focus of research on compressive failure modeling shifted to kink bands when Budiansky [44] in 1983 proposed a model for kink-band formation. He derived an expression for the compressive strength, assuming inextensible fibers, as

$$\sigma_c = G + E_T \tan \beta \tag{8.4}$$

where, E_T is the transverse Young's modulus of the UD composite and β is as defined in Figure 8.12. It is noted that when $\beta = 0$, the Budiansky prediction equals the Rosen prediction for the shear mode of fiber microbuckling.

Based on a post-buckling analysis of elastic composite laminates, Budiansky [44] argued that the compressive strength was not sensitive to defects such as fiber misalignment. However, he showed that if the composite had significant plasticity in shear, then the initial fiber misalignment had a significant effect. Based on perfect plasticity in shear, he derived the ratio of the new (plasticity-based) failure stress, σ_s, to the elasticity-based failure stress σ_c, as

$$\frac{\sigma_s}{\sigma_c} = \frac{1}{(\phi/\phi_s)+1} \tag{8.5}$$

Thus, if the misalignment angle ϕ is, say 2^0, and the shear-induced rotation, i.e., shear strain is $\phi_s = 0.002$, then the compressive failure stress is predicted to be 1/18 the elastic value σ_c.

Substituting σ_c from Equation (8.4) into Equation (8.5) and using $G = \tau_y/\phi_s$, where τ_y is the shear yield stress, one obtains,

$$\sigma_s = \frac{\tau_y + (E_T/\phi_s)\tan\beta}{\phi + \phi_s} \tag{8.6}$$

It is seen that for $\beta = 0$, the compression failure stress becomes,

$$\sigma_s = \frac{\tau_y}{\phi + \phi_s} \tag{8.7}$$

At the point when the shear-induced rotation is incipient, i.e., when $\phi_s = 0$, the elastic kink-band formation stress becomes

$$\sigma_c = \frac{\tau_y}{\phi} \tag{8.8}$$

This formula was proposed by Argon [45] in 1972 based on instability in the localized rotation of the initially misaligned fibers. In his model, the stress τ_y is the shear strength of the UD composite rather than the shear yield stress. The instability in the fiber rotation angle may be viewed as caused by the perfect (constant stress) plasticity in the shear response of the UD composite in the Budiansky model. In other words, the Argon formula is for the beginning of the kink-band formation process. In this sense, it gives a lower bound to the compression failure stress.

Note that the kink-band rotation angle β appearing in Equation (8.6) has been found by Budiansky and Fleck [46] to not have a large effect on the compression failure stress for most angles found in experimental observations. Also, the fiber misalignment angle is comparatively significantly more important.

The Argon formula for critical compressive stress, Equation (8.8), is easily derived by simple force equilibrium considerations, as depicted in Figure 8.13. Consider a region of a UD composite with fibers misaligned at an angle ϕ and subjected to axial compressive stress σ. A slice of the composite taken with an edge parallel to fibers and an edge inclined at the misalignment angle ϕ will satisfy the following force equilibrium equation along the inclined edge.

$$\sigma\left(A_\phi \sin\phi \cos\phi\right) = \tau_\phi A_\phi \tag{8.9}$$

where the normal and shear stresses on the inclined plane of area A_ϕ are σ_ϕ and τ_ϕ, respectively.

For small angles ϕ, the approximation $\sin\phi \approx \phi$ and $\cos\phi \approx 1.0$, reduces Equation (8.9) to Equation (8.8) when the shear stress τ_ϕ reaches the yield value τ_y resulting in the critical compressive stress σ_c.

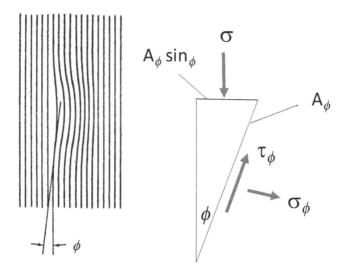

FIGURE 8.13 (left): A unidirectional composite with fibers misaligned at angle φ. (right): A slice of the composite with an inclined edge of area A_φ under axial compressive stress σ.

8.4.2.1 Defect Severity Measure for Fiber Misalignment in Wavy Fibers

As described above, the angle of fiber misalignment resulting as a manufacturing defect plays a vital role in governing the critical compressive stress to kink-band formation. In Chapter 3, Section 3.3, we described the methods for measuring the fiber misalignment such as the multiple field image analysis (MFIA) methods that use low-resolution images of cut sections of UD composites and the Fourier transform misalignment analysis (FTMA) method that uses the digitized image data transformed to the frequency domain. These methods work well for relatively small misalignment angles. Most manufacturing methods tend to result in wavy fibers when fiber bundles are used in which case larger fiber misalignment is found. For such cases, an alternative method called high-resolution misalignment analysis (HRMA) was proposed (Chapter 3, Section 3.3). Using this method, fiber misalignments in two batches of non-crimp carbon–epoxy UD composites produced with different processing parameters [47] were quantified. An example of the optical micrograph and the distribution of fiber misalignment angles is shown in Figure 8.14. The frequencies of the measured fiber misalignment angles for the two batches of composites were fit with Gaussian distributions and the 99th percentiles of the numerical values of these angles were defined as the maximum misalignment angles to be used as the ϕ-values in Equation (8.7) and (8.8) for the Budiansky and Argon models, respectively. Figure 8.15 illustrates the agreements with predictions of the two models [48]. Note that using the Budiansky model requires the measurement of the fiber rotation angle ϕ_s, which was measured separately from shear stress–strain data. As seen in Figure 8.15, the Argon model fares well for all strength data.

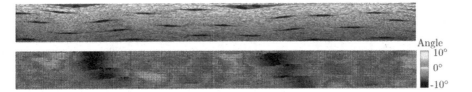

FIGURE 8.14 An optical micrograph of a section of an NCF composite (top) and the corresponding image of the section showing the fiber misalignment distribution. Source: Wilhelmsson, D., Talreja, R., Gutkin, R. and Asp, L.E., 2019. Compressive strength assessment of fibre composites based on a defect severity model. *Composites Science and Technology, 181*, 107685.

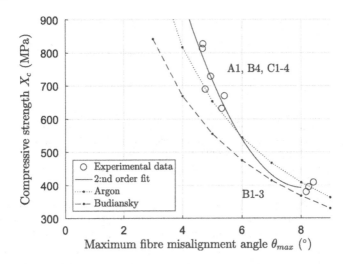

FIGURE 8.15 Experimental data of compressive strength of non-crimp carbon–epoxy composites compared with predictions of the Argon and Budiansky models using the maximum misalignment angles. Source: Wilhelmsson, D., Gutkin, R., Edgren, F. and Asp, L.E., 2018. An experimental study of fibre waviness and its effects on compressive properties of unidirectional NCF composites. *Composites Part A: Applied Science and Manufacturing, 107*, 665–674.

The uncertainty of what value to use for the misalignment angle ϕ and the observation that the fiber waviness appears as clusters of wavy fibers (Figure 8.14) prompted another approach to treating fiber misalignment for modeling purposes. That approach was described in [47] and will be summarized here.

Consider a region of a UD composite with clusters of wavy fibers, as illustrated in Figure 8.16. Under axial compressive stress σ, kink bands are likely to form within these fiber clusters. We wish to define a defect severity index S_D such that the compressive strength can be written as,

$$\sigma_c = S_D \tau_y \qquad (8.10)$$

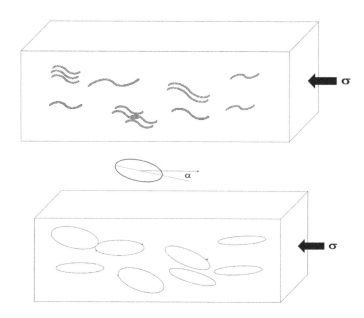

FIGURE 8.16 Top: A region of UD composite with fibers nominally oriented in the horizontal direction containing clusters of wavy fibers. Bottom: Each fiber cluster is replaced by an ellipse whose orientation is defined by the angle between the loading axis and the minor axis of the ellipse. An axial compressive stress σ acts on the composite.

This equation can be seen as a generalization of the Argon model, Equation (8.8), where for a single misalignment, $S_D = 1/\phi$. For misaligned fiber clusters, the defect severity index is developed by first replacing the fiber clusters with smeared-out regions representing the areas affected by the fiber clusters, as illustrated in Figure 8.16. The defect severity index is then assumed as a product of three "knock down" functions as,

$$S_D = f_1(A) f_2(a,\alpha) f_3(\theta) \tag{8.11}$$

where A is the area of the ellipse, a is the minor axis length of the ellipse and α is the orientation of the elliptical plane with respect to the loading axis. Finally, the angle θ represents the mean value of the fiber misalignment angles within a given cluster. The functions f_1 and f_2 were evaluated in [47] by using experimental data, and f_3 was taken as the inverse of the angle θ following the Argon model. The predictions of the kink band–induced failure stress using the defect severity index revealed that function f_3 has the most effect in determining the failure stress, while the functions f_1 and f_2 representing the size and shape of the misaligned fiber clusters have less but significant effects, f_1 more than f_2. The effect of the function f_1 on the failure stress represents the coordination of misaligned fibers within the cluster in initiating a kink band, while the function f_2, which combines the elliptical shape, i.e., the aspect ratio of the ellipse, and its inclination with respect to the potential plane of the kink band appears to have less influence than the fiber cluster size in initiating a kink band. The methodology based on the defect severity index in [47] was applied to assess the compressive strength of an aero-engine component in [49].

8.4.2.2   Effect of Matrix Voids

Matrix voids are present in a composite depending on the manufacturing process. In Chapter 3, Section 3.4, techniques for observing and quantifying matrix voids were described. It is expected that in the compression failure process, voids will play a role. Focusing on the kink-band mechanism, a study [50] found two main effects of voids: a) A kink band approaching a void tip was deflected and b) a kink band confronting a void was arrested. However, how these effects translate into a reduction of compressive strength is not clear. In an experimental study [51] of a model UD composite with a single air bubble introduced between the fibers in the curing process, the kink-band formation was observed. The observations showed fiber–matrix debonding followed by sequential fiber microbuckling prior to kink-band formation. The initial fiber misalignment was not a part of this study. In a later study [52], the compression buckling of a single fiber was analyzed by a model where the fiber was modeled as a Timoshenko beam supported on one side by the matrix, modeled as an elastic foundation, and unsupported on the other side to represent a void. The analysis indicated the role of initial fiber waviness at the void location in determining the buckling stress.

The matrix voids can be of different sizes, shapes, and spatial distribution, depending on the manufacturing process history. At the same time, the fiber waviness resulting from a manufacturing process can be of different geometrical characteristics, as described above. While the effects of the two defect types individually can be characterized, their synergistic effect on compressive stress reduction is not easy to delineate. Much more work is needed than what is currently in the literature.

8.4.3   Transverse Tension

The manufacturing defects that influence the failure of UD composites in tension transverse to fibers are a nonuniform distribution of fibers in the composite cross section and matrix voids. These defects were described in Chapter 3, Section 3.2 (nonuniform fiber distribution), and Section 3.4 (matrix voids). To a lesser degree, resin-rich areas also affect transverse failure, and these defects are described in Section 3.5. Related to the nonuniform distribution of fibers, statistical descriptors for microstructural representation were described in Chapter 4, Section 4.2, and their simulation (statistical reconstruction) was treated in Section 4.3, where quantifying the degree of fiber distribution nonuniformity was described. The concept of microstructural representation with respect to failure initiation in terms of a representative volume element (RVE) was emphasized in that chapter as it is crucial to the analysis of transverse failure. The observed mechanisms of transverse failure are described in Chapter 5, Section 5.4.

Here, we shall treat the modeling strategies for transverse failure analysis. The first step in this direction is to have a clear understanding of the failure as described in Section 5.4. The microscopy observations indicate fiber–matrix debonding as the first event of failure. It is therefore assumed in most studies that the fiber–matrix interface must be defective to cause this failure, or that failure occurs when the stress state at the fully bonded fiber–matrix interface becomes critical. However, there has been no study to show that interfacial defects initiate debonding under transverse tension.

Let us examine the assumption of failure initiating at a fully bonded interface when a critical stress state is reached. A fundamental problem with this approach is that the "strength" of the interface cannot be known. An irrefutable fact is that the stress state at any point on the fiber–matrix interface will be triaxial under uniaxial transverse tension. Expressed in terms of principal stresses, the ratios of the three principal stresses (taken as pairwise ratios or the ratio of mean to maximum) at the interface will in general be different. Thus, "strength", defined as the maximum value of the tensile stress cannot be uniquely defined. For instance, if this quantity is determined experimentally by transversely loading a single-fiber composite and recording as interfacial "strength" the largest tensile principal stress at the interface, then this value will not apply in a real composite with multiple fibers because the stress state at any point at any fiber–matrix interface will not have the same ratios of principal stresses as in the single-fiber case. This non-uniqueness of strength will also be the case if the strength is defined as the critical value of the radial tensile stress at the fiber–matrix interface. One study [53] attempted to determine which critical stress (tensile or shear) governed the failure of the fiber–matrix interface in a multifiber composite. They conducted an in-situ observation of failure initiation in a composite loaded transversely in tension and interpreted the interfacial failure stress by numerically computing the stresses on the fiber–matrix interface in a multifiber composite. Assuming the interface to have two different "strengths", one in tension and the other in shear, they focused on the radial tensile stress (called interfacial normal stress, INS in [53]) and the axial (or tangential) shear stress (called interfacial shear stress, ISS in [53]), as shown in Figure 8.17. The variation of these stresses

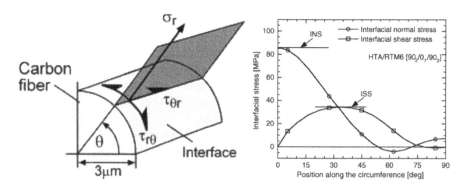

FIGURE 8.17   (a) Radial tensile and tangential (and axial) shear stresses at the fiber–matrix interface and (b): variations of these stresses along the circumference of a fiber placed in a regular pattern. Source: Hobbiebrunken, T., Hojo, M., Adachi, T., De Jong, C. and Fiedler, B., 2006. Evaluation of interfacial strength in CF/epoxies using FEM and in-situ experiments. *Composites Part A: Applied Science and Manufacturing, 37*(12), 2248–2256.

along the circumference of a fiber in a regular fiber pattern within the 90° plies of a carbon–epoxy cross-ply laminate is also depicted in Figure 8.17.

In the study [53] the authors found that in a composite where the fiber–matrix interface was "strong" (fully bonded), the failure was cohesive, i.e., the matrix close to the fiber surface failed. However, the calculated radial tensile stress to failure was significantly higher than the matrix tensile strength in the bulk (unreinforced) form and had much higher scatter than in the bulk. In contrast, the local shear stress at interfacial failure was found to be lower than the shear strength in the bulk. Such contradictory results must question the validity of the "strength" concept when the stress state at the interface is triaxial. This issue has been resolved by switching to a more fundamental approach to interfacial failure, based on energetic considerations, as described below.

8.4.3.1 Failure of Epoxies under Triaxial Tension

The effect of triaxial tension on the deformation of epoxy polymers was studied in [30]. Triaxial stresses were induced in epoxy polymers by a poker-chip test where a specimen made as a thin circular disk of large diameter (called a poker chip) was bonded on either side to aluminum rods that were then pulled in tension. Figure 8.18 compares the stress–strain response of four epoxies under uniaxial and triaxial tension [30]. What is of importance to note is that under equi-triaxial (hydrostatic) tension the deformation of epoxy loses all inelasticity, presumably due to the absence of shear.

In [28] failure initiation in epoxies under triaxial stress states was examined with energy considerations. Recognizing that the strain energy density at any point is a sum of its volume expansion-related (dilatational) part and shear-driven distortional part, the initiation of failure was viewed as a competition between two mechanisms:

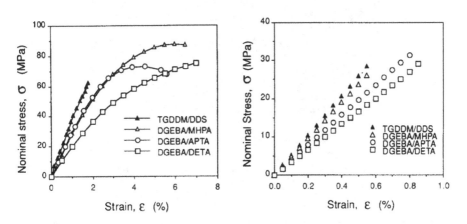

FIGURE 8.18 Nominal stress-strain response of four epoxies under (a) uniaxial tension and (b) equi-triaxial tension. Source: Asp, L.E., Berglund, L.A. and Gudmundson, P., 1995. Effects of a composite-like stress state on the fracture of epoxies. *Composites science and technology, 53*(1), 27–37.

Microcavitation caused by volume expansion and yielding induced by distortion. Depending on the triaxiality of the stress state and the threshold values of the dilatational and distortional energy densities, one of the two mechanisms will reach first. For three epoxies the two critical values were determined experimentally by tests consisting of uniaxial and biaxial tension tests, thermally loaded disk test, and poker-chip test. While the critical dilatational and distortional energy density values differed as measured by different methods, the critical value of the former was always found to be lower than that of the latter, suggesting that if hydrostatic tension is present, it will always initiate microcavitation before yielding occurs. This important result questions many failure analyses reported in the literature where the criticality of the dilatational energy density is not taken into consideration.

The mechanism of dilatation-induced cavitation was further studied by molecular dynamic simulation in [31]. Using the molecular structure of EPON-862+DETDA epoxy resin, it was shown that the stress–strain response depended significantly on the macroscopic stress state. By studying the behavior under uniaxial, equi-biaxial, and equi-triaxial tension it was shown that void formation was significantly higher under imposed equi-triaxial tension than under the other two stress states. These findings gave further credence to the mechanics-based interpretation of dilatation energy density as the cause of brittle cavitation.

The mechanism of microcavitation is difficult to observe directly as its size is of molecular dimensions but its consequence on a fracture surface can be seen, as shown in Figure 8.19 [28], which is an SEM image of the fracture surface of a

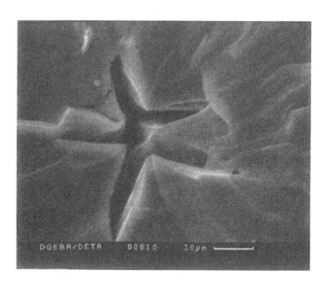

FIGURE 8.19 Scanning electron micrograph of the fracture surface of a poker-chip specimen of DGEBA/DETA showing the initiation of brittle cavitation. Source: Asp, L.E., Berglund, L.A. and Talreja, R., 1996. A criterion for crack initiation in glassy polymers subjected to a composite-like stress state. *Composites Science and Technology*, *56*(11), 1291–1301.

poker-chip specimen at the fracture initiation site. The radial direction of fracture propagation suggests that equal tension in all radial directions operated at the instant of failure initiation. The absence of shear provides brittle conditions (Figure 8.18) in which the cavity expansion becomes unstable. Figure 8.20(a) schematically depicts the expansion of a cavity formed under equi-triaxial tension and Figure 8.20 (b) illustrates that if a slight bias existed in the principal stresses, expressed by one of the principal stresses being larger by the value $\Delta\sigma$ than the other principal stresses, then a crack forms normal to that principal stress [28].

8.4.3.2 Initiation of Failure: Uniform Fiber Distribution

In a study [29] the authors examined the occurrence of microcavitation versus yielding in the epoxy matrix within a UD composite under transverse tension. Assuming a uniform distribution of fibers in the composite cross section they calculated the local dilatational and distortional energy density values under a thermal cooldown from the curing temperature and imposed transverse tension. Considering three fiber arrangement patterns – square, hexagonal, and square-diagonal – with fiber volume fractions varying between 0.2 and 0.7 or 0.8, the location of maximum dilatational energy density was consistently found to be close to the fiber surfaces, while where yielding initiated in the matrix varied depending on the fiber packing arrangement. For one case of square fiber array, Figure 8.21 schematically illustrates the initiation sites for microcavitation and yielding.

At this point, a cautionary note regarding the dilatational energy density criterion for the initiation of brittle cavitation in glassy polymers is in order. In the literature,

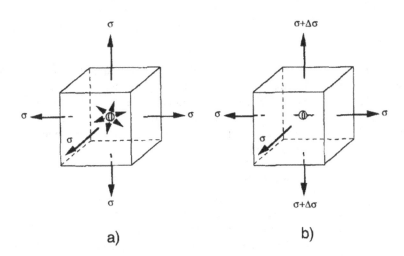

a)   b)

FIGURE 8.20 (a) Nucleation and equiaxial growth of a cavity under eqi-triaxial tension. (b) Crack formation and growth in a preferred direction are normal to the maximum principal stress. Source: Asp, L.E., Berglund, L.A. and Talreja, R., 1996. A criterion for crack initiation in glassy polymers subjected to a composite-like stress state. *Composites Science and Technology*, 56(11), 1291–1301.

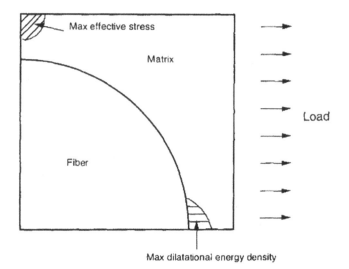

FIGURE 8.21 The locations of the maximum values of dilatational energy density and distortional energy density (same as von Mises effective stress) in a square fiber pattern. Source: Asp, L.E., Berglund, L.A. and Talreja, R., 1996. Prediction of matrix-initiated transverse failure in polymer composites. *Composites Science and Technology*, 56(9), 1089–1097.

it is sometimes assumed that this criterion can also be expressed in terms of the hydrostatic stress. This can lead to error. The criterion is as follows,

$$U_v = \frac{1-2\upsilon}{6E}(\sigma_1 + \sigma_2 + \sigma_3)^2 = U_v^{crit} \qquad (8.12)$$

where σ_1, σ_2, and σ_3 are the principal stresses, which are equal for the hydrostatic stress state, and E and υ are Young's modulus and Poisson's ratio, respectively. As shown in [28], the value of the critical dilatational energy density was found to be the same at two different temperatures. Expressing the criterion in terms of the mean of stress will require assuming that the two elasticity constants are independent of temperature. This is generally not the case for polymers.

8.4.3.3   Initiation of Failure: Nonuniform Fiber Distribution

In real composites, the fibers in a UD composite are nonuniformly distributed. This manufacturing defect was described in Chapter 3, Section 3.2 where methods for characterizing the nonuniformity using inter-fiber distance statistics were discussed. In Chapter 4, Section 4.2 the representation of microstructure with a representative volume element (RVE) was discussed and in Section 4.3 simulation of microstructure using the statistical descriptors was described. As discussed there, while methods for constructing RVEs commonly represent the randomness of nonuniform fiber distribution, only approaches that use fiber mobility and fiber clustering in the statistical reconstruction of fiber distribution achieve quantification of the degree of nonuniformity in the distributions. This aspect is important if the objective of failure analysis is eventually to control manufacturing processes to improve the performance of composite materials. In the following the first event of failure in UD composites having nonuniform fiber distribution will be described and the results quantified with two approaches: a) In terms of the degree of nonuniformity of fiber distribution, b) in terms of fiber mobility to capture fiber clustering. The following sections will treat these separately.

8.4.3.3.1 Degree of Fiber Distribution Nonuniformity

The first study to quantify the defect severity in the case of nonuniform fiber distribution in a UD composite cross section was [54, 55], as described in Chapter 4, Section 4.3.4.1. Based on this approach, an RVE realization for a given fiber diameter, fiber volume fraction, and degree of fiber distribution nonuniformity was generated. An algorithm developed for this purpose is described in Figure 8.22. An RVE realization was then subjected to transverse loading as depicted in Figure 8.23.

The local stress states in the matrix for five or more realizations were computed by the finite element method for each case of three different fiber volume fractions and three different degrees of fiber distribution nonuniformity. Figure 8.24 illustrates one case of glass–epoxy of 40% fiber volume fraction where 60% and 100% degree of nonuniformity cases are compared with the case of uniform fiber distribution. At the imposed transverse strain of 0.4% after a thermal cooldown of 82°C, the dilatational energy density values are calculated, and their contours are shown in the figure. As seen there, the maximum value of the dilatational energy density occurs at all polar points, i.e., intersections of the loading axis with the fiber surfaces, in the uniform fiber distribution case, while for nonuniform distribution cases, the location of maximum values are found to depend on the degree of nonuniformity. The energy density values increase near fiber surfaces as the inter-fiber distance reduces.

In the study [55], for each RVE realization of specified fiber volume fraction and degree of nonuniformity of fiber distribution, a finite element mesh was generated using the software ABAQUS, and after a convergence study, a search was conducted to find the point where the largest hydrostatic tension occurred. The pairwise ratios of the three principal stresses, and the ratio of the maximum principal stress to the average principal stress, were calculated. This was done at each increment of 0.1% of the imposed transverse strain. A typical plot showing the results for 50% fiber volume fraction and 100% degree of fiber distribution nonuniformity is displayed in Figure 8.25.

As illustrated by this plot, at the end of the thermal cooldown, and before the transverse mechanical loading, the three principal stresses are quite unequal. As the transverse loading increases from zero, they begin to become less unequal, and at a certain imposed strain they are least unequal, following which point, they diverge. This pattern is seen as a typical behavior. Thus, as the transverse loading is applied, the shear deformation at the point of the largest principal stress within the RVE reduces and the conditions for brittle cavitations improve. If the three principal stresses become equal, then the conditions for brittle cavitation would be ideal,

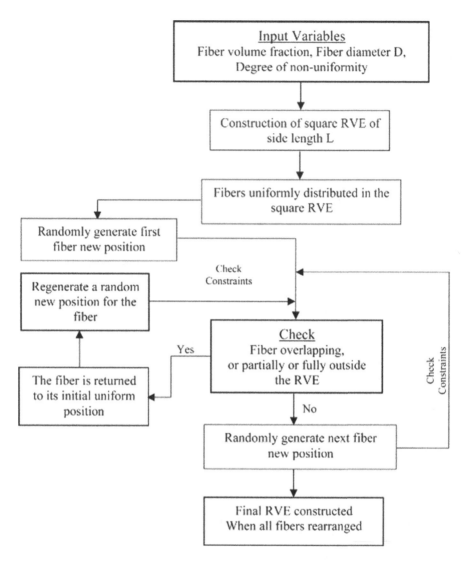

FIGURE 8.22 Flowchart of the algorithm for constructing an RVE realization. Source: Elnekhaily, S.A. and Talreja, R., 2018. Damage initiation in unidirectional fiber composites with different degrees of nonuniform fiber distribution. *Composites Science and Technology*, *155*, 22–32.

provided the dilatational energy density at that point equals or exceeds the critical value (~ 0.2 MPa for epoxies [54]).

In Figure 8.26 the dilatational and distortional energy densities corresponding to the data on principal stresses in Figure 8.25 are plotted. As seen in the figure, the two energy densities increase differently with the applied strain. The dilatational energy density attains its critical value of ~0.2 MPa at approximately 0.3% strain at which

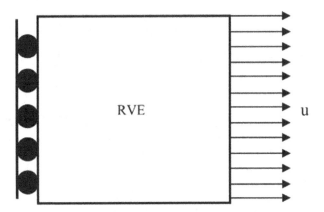

FIGURE 8.23 An RVE realization subjected to uniform transverse displacement. Source: Elnekhaily, S.A. and Talreja, R., 2018. Damage initiation in unidirectional fiber composites with different degrees of nonuniform fiber distribution. *Composites Science and Technology, 155*, 22–32.

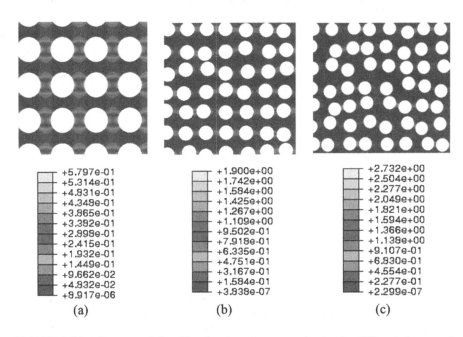

FIGURE 8.24 Contours of the dilatational strain energy density for different degrees of nonuniformity (a) 0% nonuniformity, (b) 60% nonuniformity, and (c) 100% nonuniformity for 40% fiber volume fraction under 0.4% mechanical strain. Source: Elnekhaily, S.A. and Talreja, R., 2018. Damage initiation in unidirectional fiber composites with different degrees of nonuniform fiber distribution. *Composites Science and Technology, 155*, 22–32.

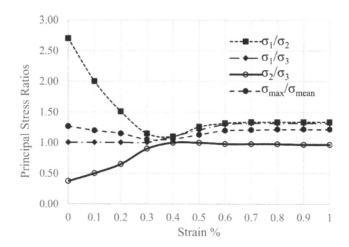

FIGURE 8.25 The principal stress ratios and the ratio of maximum/mean principal stress versus applied strain for 50% fiber volume fraction and 100% degree of nonuniformity. Source: Elnekhaily, S.A. and Talreja, R., 2018. Damage initiation in unidirectional fiber composites with different degrees of nonuniform fiber distribution. *Composites Science and Technology, 155*, 22–32.

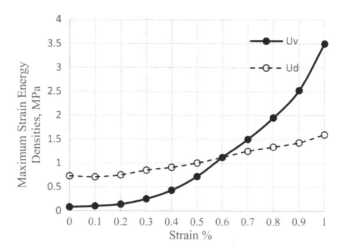

FIGURE 8.26 The maximum values of dilatational energy density U_v and distortional energy density U_d attained within RVEs versus applied strain for 50% fiber volume fraction and 100% degree of nonuniformity corresponding to data in Figure 24. Source: Elnekhaily, S.A. and Talreja, R., 2018. Damage initiation in unidirectional fiber composites with different degrees of nonuniform fiber distribution. *Composites Science and Technology, 155*, 22–32.

value the principal stresses are also equal within the tolerance of 20%. Figure 8.26 also shows that the distortional energy density U_d does not reach its critical value of ~1.0 MPa until 0.5% strain. This behavior was found in all cases examined suggesting that brittle cavitation always occurs before yielding initiates within the epoxy matrix.

The results of the study [55] are summarized in Figure 8.27 where the applied strain at which brittle cavitation occurs is plotted against the degree of fiber distribution nonuniformity for three different fiber volume fractions. The curves shown are drawn through the averages of the initiation strains taken over five RVE realizations and the scatter in the results is also indicated. As the figure shows, the strains are the lowest for 60% fiber volume fraction among the three cases as the fibers are spaced closest on average for this case. Also, as the degree of nonuniformity increases, more fibers occupy spaces closer to their neighbors than in the uniform distribution pattern for the same fiber volume fraction. This results in lowering of the strain for the onset of brittle cavitation. It is noted that the results shown in Figure 8.27 are the first of their kind as they quantify the effects of fiber distribution nonuniformity with well-defined defect severity metrics.

8.4.3.3.2   Fiber Mobility-Based Fiber Clustering

The RVE realizations based on fiber mobility simulation described in Chapter 4, Section 4.3.4.2, result in fiber distribution patterns such as those in Figure 4.9 of Chapter 4. These patterns are not amenable to the square-shaped RVEs for stress analysis using finite element methods where uniform boundary conditions are specified on straight boundaries as in Figure 8.23. An embedded cell model is used instead [56] following [57]. The cell is defined by the circular region (RVE) having a minimum radius R that encircles the fibers. As shown in Figure 8.28(a), this circle is embedded within the outer region of a homogenized composite having the same fiber volume fraction as the cell. The size of the outer region of side length L is determined using Hill's concept [13] of minimum RVE size for effective elastic

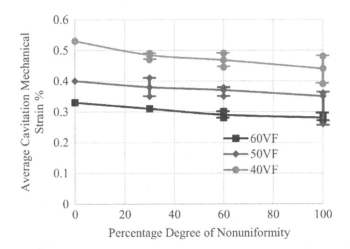

FIGURE 8.27   The average mechanical strain at the onset of cavitation versus the percent degree of nonuniformity for 40, 50, and 60% fiber volume fraction. The range of variation over five RVE realizations is indicated. Source: Elnekhaily, S.A. and Talreja, R., 2018. Damage initiation in unidirectional fiber composites with different degrees of nonuniform fiber distribution. *Composites Science and Technology, 155*, 22–32.

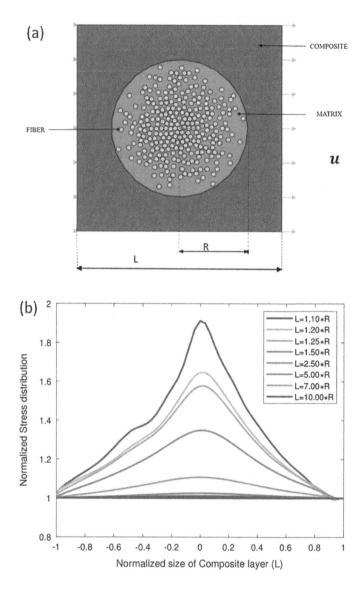

FIGURE 8.28 (a) A local region of nonuniform fiber distribution embedded as a cell within a square outer region of homogenized composite and (b) variation of the normal stress on the boundary, normalized by its average value, with the length L normalized by the embedded cell radius R. Source: Sudhir, A. and Talreja, R., 2019. Simulation of manufacturing induced fiber clustering and matrix voids and their effect on transverse crack formation in unidirectional composites. *Composites Part A: Applied Science and Manufacturing, 127*, 105620.

properties. Accordingly, for a uniform displacement u applied to the vertical boundary of the embedded cell model the dimension L is increased until the traction on this boundary becomes uniform, see Figure 8.28(b).

As in the other case of square-shaped RVE for nonuniform fiber distribution, a thermal cooldown of $\Delta T = -82°C$ was also used here before imposing the tensile displacement. Five realizations of each RVE defined by the mobility parameter pair (dr, dθ) were generated as described in Chapter 4, Section 4.3.4.2, and were used as embedded cells. Figure 8.29 illustrates one case of the maximum dilatational energy density among all nodes in the matrix FE mesh within the cell as the mechanical strain is increased from zero. For this case of dr = 0.5r and dθ = 15°, the maximum dilatational energy density meets the critical value of 0.17 MPa for the epoxy used at the mechanical strain of ~ 0.44%. The three principal stresses at the point in the matrix where this condition was met were found to be nearly equal, with the ratio of mean to max values of ~0.95, assuring the presence of hydrostatic tension. The occurrence of the brittle cavitation condition was found to depend on the radial mobility parameter k (dr = kr) but was relatively insensitive to the dθ values.

Figure 8.30 shows the transverse tensile strain at which brittle cavitation occurs as the radial mobility parameter dr is increased for a constant dθ of 15°. The strain at which yielding initiates according to the distortional energy density criterion (same

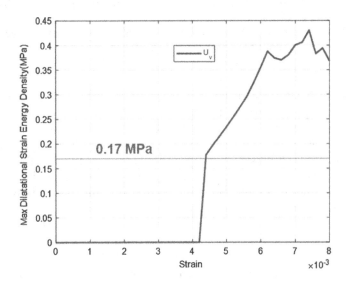

FIGURE 8.29 Variation of the maximum dilatational energy density in the matrix within RVE as the mechanical strain on the RVE is increased. Source: Sudhir, A. and Talreja, R., 2019. Simulation of manufacturing induced fiber clustering and matrix voids and their effect on transverse crack formation in unidirectional composites. *Composites Part A: Applied Science and Manufacturing, 127*, 105620.

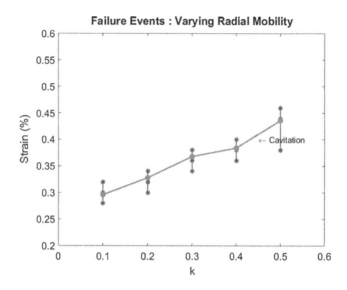

FIGURE 8.30 Transverse strain to initiation of brittle cavitation at different radial mobility levels, k. Source: Sudhir, A. and Talreja, R., 2019. Simulation of manufacturing induced fiber clustering and matrix voids and their effect on transverse crack formation in unidirectional composites. *Composites Part A: Applied Science and Manufacturing, 127,* 105620.

as the v. Mises yield criterion) was found to be higher in each case, suggesting that brittle cavitation is more likely to be the first failure event.

8.4.3.3.2.1 Initiation of debonding

As described above, the dilatation-induced brittle cavitation occurs close to fiber surfaces in an epoxy matrix under transverse tension. Also, as discussed above in Section 8.4.3.1, the criterion for this failure event has been established to be given by the critical value of the dilatational energy density, which can be found by testing the epoxy under hydrostatic tension in an independent test and is thus a material constant. As illustrated schematically in Figure 8.20, if a slight bias in the principal stresses exists toward one direction, then it is likely that the spherical cavity formed by volume expansion will grow unstably forming a brittle crack normal to that direction. Now, since the likely location for the brittle cavitation is close to a fiber surface, and the radial tensile stress is a principal stress (Figure 8.17), it is highly likely that the brittle crack formed by unstable cavity growth will lie tangentially to the fiber surface. Such a crack can be safely assumed as the fiber–matrix interface crack, i.e., a debond.

The hypothesis just described above gives a strong basis for assuming the initiation site for fiber–matrix debonding as the point of brittle cavitation but leaves the question open as to how large the initiated debond crack will be. However, the size of the initial debond turns out not to be of significance if it is assumed to be small compared to the radius of the fiber. This will become clear as we treat the formation of a transverse crack next.

8.4.3.4   Transverse Crack Formation

Figure 8.31 illustrates the progression in the number of sites within an RVE where the brittle cavitation criterion is met. As seen in the figure, at the applied strain of 0.38% the cavitation sites are few and are spread out. With a slight increase of strain to 0.4%, the cavitation occurs at more sites and these sites tend to align themselves along the vertical direction, i.e., normal to the applied load direction. With a further increase in the applied strain to 0.42%, the number of cavitation sites increases drastically, and a pattern is not easy to find.

The observed transverse cracks are shown in Figure 8.32 [58, 59]. These images suggest that the coalescence of fiber–matrix debonds leads to the formation of transverse cracks. Clearly, a minimum of two adjacent debonds must coalesce for the shortest transverse crack to form. Thus, in Figure 8.33 the strains at which two adjacent brittle cavitation sites were found in the RVEs corresponding to different fiber mobility values are plotted along with their values at the onset of brittle cavitation. Experimental values of strains to transverse failure in a glass–epoxy UD composite at two fiber volume fractions reported in the literature [60] are also shown in the figure.

The procedure for determining the strain at which a transverse crack forms, as described above, does not account for the linkup of debond cracks as suggested by the images in Figure 8.32. Thus, the strain determined this way can only be viewed as an estimate. Several studies have been conducted to analyze the linkup process, which consists of a debond crack growth on the fiber–matrix interface and its kink out into the matrix. The findings of these studies will be reviewed next.

8.4.3.4.1   Debond Growth and Kink Out

Before addressing kink out of a debond crack from the interface into the matrix one must analyze the debond growth on the interface. For this purpose, a single-fiber

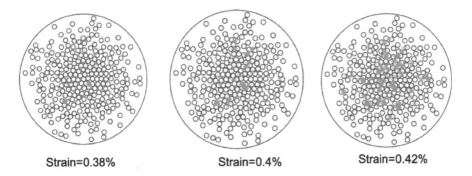

Strain=0.38%   Strain=0.4%   Strain=0.42%

FIGURE 8.31 Increasing number of cavitation sites as the applied transverse strain is increased. The tensile displacement applied is in the horizontal direction. Source: Sudhir, A. and Talreja, R., 2019. Simulation of manufacturing induced fiber clustering and matrix voids and their effect on transverse crack formation in unidirectional composites. *Composites Part A: Applied Science and Manufacturing, 127,* 105620.

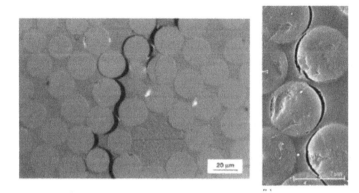

FIGURE 8.32 Transverse cracks observed in (a) glass–epoxy and (b) carbon–epoxy laminates. Source: (a) Gamstedt, E.K. and Sjögren, B.A., 1999. Micromechanisms in tension-compression fatigue of composite laminates containing transverse plies. *Composites Science and Technology*, 59(2), 167–178. (b): Romanov, V.S., Lomov, S.V., Verpoest, I. and Gorbatikh, L., 2015. Modelling evidence of stress concentration mitigation at the micro-scale in polymer composites by the addition of carbon nanotubes. *Carbon*, 82, 184–194.

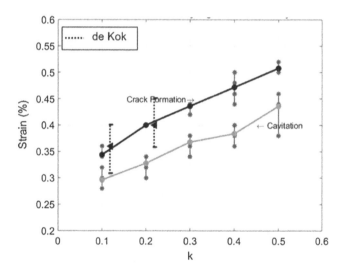

FIGURE 8.33 The transverse strain at the formation of a crack by coalescence of two neighboring cavitation (debond) sites (black circles) at different fiber mobility levels. The red circles are repeated from Figure 8.30. Experimentally measured strains to transverse failure at two fiber volume fractions reported in de Kok [60] are also indicated. Source: Sudhir, A. and Talreja, R., 2019. Simulation of manufacturing induced fiber clustering and matrix voids and their effect on transverse crack formation in unidirectional composites. *Composites Part A: Applied Science and Manufacturing*, 127, 105620. Test data from De Kok, J.M.M. and Meijer, H.E.H., 1999. Deformation, yield and fracture of unidirectional composites in transverse loading: 1. Influence of fibre volume fraction and test-temperature. *Composites Part A: Applied Science and Manufacturing*, 30(7), 905–916.

composite was considered in an analytical study [61] and in several numerical studies (e.g., [62–64]). The kink out of an interface crack in a single-fiber composite was also examined in [65]. Recognizing the importance of interaction between fibers, a later study considered a two-fiber composite [66] and discovered the effect of the relative positioning of the two fibers with respect to the loading axis. Clearly, in a multifiber composite more such interactions take place. Several studies analyzed these interactions using established fracture mechanics methods such as energy release rates (ERRs) [67, 68]. Other approaches include debond "damage" growth by cohesive zone models (e.g., [69]). As discussed in [67], cohesive zone models require interface properties that cannot be determined independently, and their application can therefore lead to questionable results. In a fracture mechanics approach taken in earlier works [62–66], and followed in [67, 68], the debond growth and kink-out analyses do not require interface properties, only the variations of ERRs along the fiber surface. These studies will be reviewed in the following.

Figure 8.34 describes the composite configuration analyzed in [67, 68]. The fibers are placed in a hexagonal pattern with a central fiber that is taken to have debonded. The debond crack is placed on one side of the central fiber following a previous study that had shown that one debond rather than two symmetrically placed debonds on a fiber surface under transverse tension was energetically more favorable [64]. The discrete multifiber region is embedded in a homogenized composite region to ensure proper transfer of boundary load to the region of interest. The embedded cell size is selected to ensure that the fiber volume in the cell is the same as that of the composite. The inter-fiber distance in the hexagonal fiber pattern is decreased as a parameter, which increases the local fiber volume fraction. The debond tip region is shown in detail in Figure 8.34(b). The incremental debond crack length as shown there is the arc length $r_f \cdot d\theta$, which is used in calculating the ERRs by the virtual crack closure technique (VCCT). The variations of mode I, mode II, and total ERRs are shown in Figure 8.35 for two different inter-fiber distances normalized by the fiber radius ($IDn = ID/r_f$) in the hexagonal fiber pattern. The ERRs are normalized by G_0, which is a reference value corresponding to a debond crack in an isotropic medium.

Certain characteristic features of the debond crack growth can be noted in Figure 8.35. As the crack progresses on the fiber–matrix interface, the ERRs increase first and then decrease. The mode I ERR achieves a peak value at small debond angles, while the mode II ERR peaks much later. Also, reducing the inter-fiber spacing has a larger effect on the mode II ERR peak value. On including the thermal cooldown of 100°C the trends in the variations in the ERRs seen in Figure 8.35 at a mechanical strain of 0.5% did not change significantly. The compressive radial stress induced by the thermal cooldown causes a reduction in mode I ERR and once the debond surfaces close at $\theta = 70°$, it increases the size of the contact between the debonded surfaces.

The debond kink out in the matrix was studied by including the kinking of the debond crack in the finite element model, as displayed in Figure 8.36 [67]. For each debond angle θ, a kinked crack of a short length L = $0.04r_f$ was assumed to emanate from the debond tip and grow into the matrix at an angle θ_k with respect to the horizontal axis of the model. This angle was determined based on the maximum

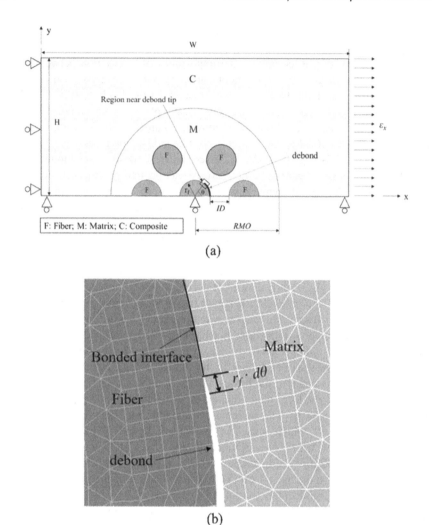

FIGURE 8.34 (a) A multifiber composite model for analyzing fiber–matrix debond growth and (b) the region around the debond crack tip. Due to symmetry, only half of the composite region is used in a finite element analysis. Source: Zhuang, L., Talreja, R. and Varna, J., 2018. Transverse crack formation in unidirectional composites by linking of fibre/matrix debond cracks. *Composites Part A: Applied Science and Manufacturing, 107*, 294–303.

energy release rate criterion, i.e., the crack kinks out of the interface in a direction that maximizes the total ERR of the kinked crack. The variations of ERRs for one normalized inter-fiber distance and one fiber volume fraction for the cases of debond with and without kinking are shown in Figure 37. As seen in this figure, the mode II ERR falls to nearly zero value for kinked debond cracks at all debond angles, suggesting that the kinked crack tip grows in the opening mode. The largest ERR

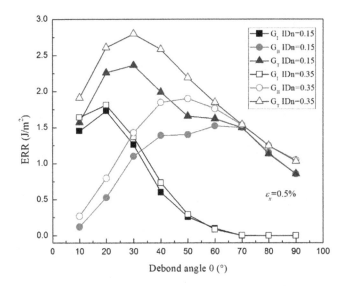

FIGURE 8.35 The variation of ERRs with the debond angle q for the model is shown in Figure 8.34 at the composite fiber volume fraction $V_f = 0.6$ and imposed strain $\varepsilon_x = 0.5\%$. The mode I, mode II, and total ERRs corresponding to two different inter-fiber distance *ID* are indicated. Source: Zhuang, L., Talreja, R. and Varna, J., 2018. Transverse crack formation in unidirectional composites by linking of fibre/matrix debond cracks. *Composites Part A: Applied Science and Manufacturing, 107,* 294–303.

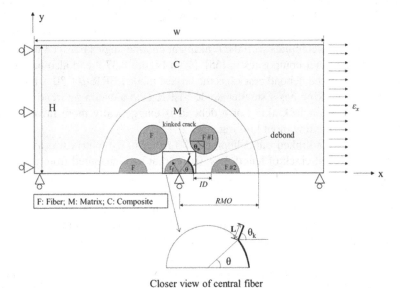

Closer view of central fiber

FIGURE 8.36 The embedded cell model in which a debond crack of a short length L is taken to kink out at the angle θ_k. Source: Zhuang, L., Talreja, R. and Varna, J., 2018. Transverse crack formation in unidirectional composites by linking of fibre/matrix debond cracks. *Composites Part A: Applied Science and Manufacturing, 107,* 294–303.

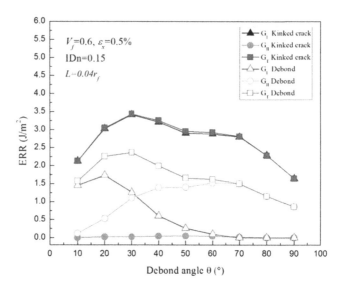

FIGURE 8.37   Variation of ERRs for debond and kinked cracks at *IDn* = 0.15. Source: Zhuang, L., Talreja, R. and Varna, J., 2018. Transverse crack formation in unidirectional composites by linking of fibre/matrix debond cracks. *Composites Part A: Applied Science and Manufacturing, 107*, 294–303.

among all possible kinked cracks occurs at $\theta \approx 30°$ for the normalized inter-fiber spacing *IDn* = 0.15. This angle was found to be $\theta \approx 40°$ for a higher *IDn* = 0.35. Thus, increasing the inter-fiber distance increases the debond angle where kinking occurs. This result complies with the kinking at debond angles between 60° and 70° found for single-fiber composites in [65]. From Figure 8.37 it can also be observed that when the kinked debond crack has the largest mode I ERR ($\theta \approx 30°$), the debond crack without kinking has a smaller mode I ERR with a declining trend. This suggests that at that angle kinking of a debond is energetically more favorable than continuing to debond without kinking.

   The growth of a kinked-out debond crack further in the matrix was also studied in [67]. The kinked crack of length L = $0.04r_f$ that had emanated from the debond crack at the interface was regarded as the initial matrix crack and its further growth was evaluated. An increment of the same length L was added to the kinked crack and its direction was determined as that which maximized the total ERR, in the same way as the direction of the initial kinked crack was determined. This procedure was repeated for one more crack increment. Like the initial kinked crack, the further crack growth was also found to be dominated by mode I ERR. The crack path was calculated for the two inter-fiber spacing values considered and is displayed in Figure 38 for *IDn* = 0.35. As seen in the figure, the crack path is affected by the neighboring fiber labeled Fiber #1, particularly when the crack tip approaches this fiber. The angle of approach of the crack is not directed at the fiber but is deflected by the local stress field surrounding the fiber.

The interaction of the matrix crack with the fiber in its path was also examined in [67]. As the kinked crack advances toward Fibre #1 (Figure 8.38), it produces a stress concentration on the surface of that fiber. This could result in the failure of the fiber–matrix interface if the stress state reaches the critical condition needed for interface failure to occur. Assuming the radial tensile stress at the interface to govern the interface failure, this stress normalized by the remotely applied stress on the composite (at $\varepsilon_x = 0.5\%$) was calculated and is displayed in Figure 8.39 for Fiber #1. The position along the circumference of the fiber is given by the angle θ_n measured from the vertical direction (see Figure 8.36), and positive going clockwise. The stress variation shown in the figure is for the inter-fiber spacing $IDn = 0.35$ for which the crack kinking angle $\theta = 40°$ was taken (as discussed above). The figure shows the stress variation for two kinked crack lengths, the initial (first) and the last (third), as shown in Figure 8.38. For the initial kinked crack length, when the crack tip is relatively far from the surface of the neighboring fiber, the radial tensile stress variation shows two peaks. As the kinked crack tip approaches the fiber surface, a sharp increase in the stress peak closest to the crack tip is found. This suggests strongly that the fiber–matrix interface is likely to fail at that point. The ligament of the matrix material between the kinked crack tip and the debond of Fiber #1 will then also fail, leading to the growth of the kinked crack into the interface.

The debonding of a neighboring fiber due to an approaching kinked-out crack described above has characteristic features that are expected to hold for all multiple fiber composites. Also, the shielding effect of fibers on the ERRs of debond cracks, illustrated in Figure 8.35 for two inter-fiber distances, suggests that the transverse crack formation will require higher loading as the fibers come closer.

In an experimental study [69] of transverse crack formation it was found that depending on the polymer matrix the curing process can result in multiple discrete fiber–matrix debond cracks. Transverse cracks were then formed on mechanical

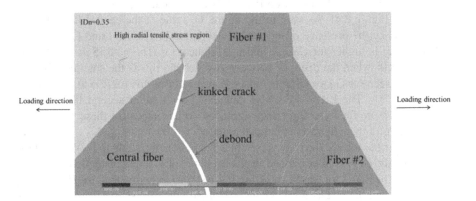

FIGURE 8.38 Kinked crack path in a matrix and radial stress along the Fiber#1 interface. ($IDn = 0.35°, \theta = 40°$). Source: Zhuang, L., Talreja, R. and Varna, J., 2018. Transverse crack formation in unidirectional composites by linking of fibre/matrix debond cracks. *Composites Part A: Applied Science and Manufacturing, 107,* 294–303.

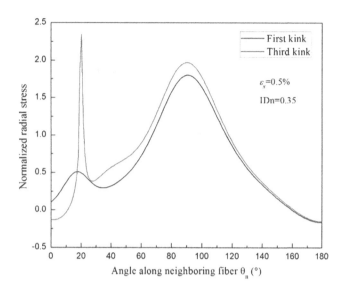

FIGURE 8.39   Radial stress variation along the interface of Fiber#1 at two kinked crack lengths. Source: Zhuang, L., Talreja, R. and Varna, J., 2018. Transverse crack formation in unidirectional composites by linking of fibre/matrix debond cracks. *Composites Part A: Applied Science and Manufacturing*, 107, 294–303.

loading by linking these cracks. This scenario was also analyzed in [67] by considering a debond crack on the central fiber in the hexagonal fiber cluster as before but with an additional debond crack on a neighboring fiber in two cases, as shown in Figure 8.40. The difference between the two cases is that in Case I the neighboring debond is closer to the central debond than in Case II. The differences in the two cases in terms of the effect on the ERRs of the central fiber debond crack are depicted in Figure 8.41. As seen in the figure, G_I value of the central fiber debond is enhanced by the presence of the second debond in both Case I and Case II. The enhancement is higher for Case I, and it is highest for $IDn = 0.15$ at the debond angle $\theta = 60°$ when the distance between two debond tips is the smallest. Similar behavior is seen for G_{II}, except the most enhancement in this ERR is at the debond angle $\theta = 70°$. The ERRs were also calculated for smaller second fiber debond lengths and were found to show less significant enhancement in these values than those shown for the neighboring debond angle $\theta' = 60°$.

Going further toward the transverse crack formation, the study [67] examined the kinking of the central fiber debond in the presence of a neighboring debond in the two cases. The ERRs for Case I and Case II are shown in Figure 8.42 for $IDn = 0.15$ and 0.35, where the results for a single debond case are also shown for comparison. For Case I, the G_I value of the kinked crack is enhanced most at the debond angle $\theta \approx 60°$ for $IDn = 0.15$. The same is seen for Case II, except at a higher debond angle. Comparing these results to those for the kinked crack without debonding of the neighboring fiber, we see that the presence of the second debond enhances

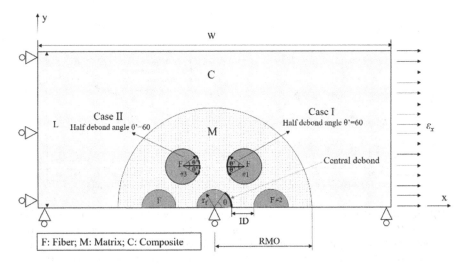

FIGURE 8.40 A debond on the central fiber with two cases of neighboring debond cracks. Source: Zhuang, L., Talreja, R. and Varna, J., 2018. Transverse crack formation in unidirectional composites by linking of fibre/matrix debond cracks. *Composites Part A: Applied Science and Manufacturing*, *107*, 294–303.

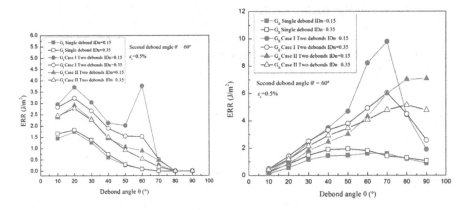

FIGURE 8.41 The ERRs of the central debond crack with and without the presence of a neighboring debond in Case I and Case II. Left: Mode I ERRs, and right: mode II ERR. Source: Zhuang, L., Talreja, R. and Varna, J., 2018. Transverse crack formation in unidirectional composites by linking of fibre/matrix debond cracks. *Composites Part A: Applied Science and Manufacturing*, *107*, 294–303.

the ERRs significantly. The debond angle θ at which the most enhancement occurs is, however, shifted to higher angles, reflecting the effect of this enhancement of the proximity of the second debond crack tip. As in the case of the debond growth described above, Case I turns out to be more favorable than Case II for crack kinking, as evidenced by the ERR values.

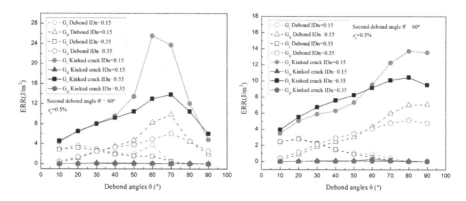

FIGURE 8.42 Variations of ERRs for the central fiber kinked debond at two inter-fiber distances *IDn* for Case II (left) and Case II (right). Source: Zhuang, L., Talreja, R. and Varna, J., 2018. Transverse crack formation in unidirectional composites by linking of fibre/matrix debond cracks. *Composites Part A: Applied Science and Manufacturing*, *107*, 294–303.

Taking $\theta = 60°$ as the angle at which the debond of the central fiber kinks out, its further growth in the matrix was then examined under the influence of two different debond lengths of Fiber #1, i.e., two variations of Case I. It was found that the mode I ERR of the kinked crack increases with its length for $\theta' = 60°$, while it decreases for the shorter debond at $\theta' = 30°$. This suggests that when the second debond crack tip is closer to the kinked crack, it tends to intensify the growth rate of the kinked crack. On the other hand, when the neighboring fiber is fully bonded or insufficiently debonded, it dampens the growth of the kinked crack.

8.4.3.4.2 Effects of Matrix Voids

Matrix voids were described in Chapter 3, Section 3.4. The effects of voids on transverse crack formation have been studied extensively. Very clear evidence of the effect of a void in a resin-rich layer of a UD composite under transverse tension was reported in [70] and is reproduced in Figure 8.43. The process of crack initiation appears to be influenced by the presence of a void. It is therefore of interest to begin studying how the first event in the transverse cracking process is affected by voids. In a study of the effect of matrix voids on brittle cavitation [56] voids were placed in the matrix in the inter-fiber regions within an RVE. The local stress field generated by nonuniformly distributed fibers within the RVE is perturbed by the presence of voids. The enhancement of the local principal stresses responsible for cavitation depends on the size and shape of a void as well as its position (distance from fibers). As the void volume fraction increases, more voids of a given void size occupy positions between fibers, increasing the probability of reaching the critical dilatation energy density needed for cavitation. Figure 8.44 shows the effect of increasing the void volume fraction while keeping the void size constant at 0.1 times the fiber diameter. A significant earlier cavitation is predicted. A similar effect was found if the void volume fraction was kept constant and the void size was increased [56, 71].

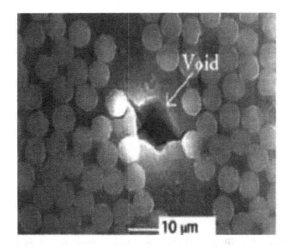

FIGURE 8.43   A matrix void inducing fiber–matrix debonding in a local region of nonuniformly distributed fibers. Source: Wood, C.A. and Bradley, W.L., 1997. Determination of the effect of seawater on the interfacial strength of an interlayer E-glass/graphite/epoxy composite by in situ observation of transverse cracking in an environmental SEM. *Composites science and Technology*, *57*(8), 1033–1043.

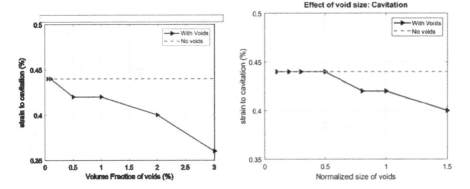

FIGURE 8.44   Strain to brittle cavitation at increasing void volume fraction for voids of 0.1× the fiber diameter, left, and at increasing void size for void volume fraction of 0.1%, right. Source: (Left) Sudhir, A. and Talreja, R., 2019. Simulation of manufacturing induced fiber clustering and matrix voids and their effect on transverse crack formation in unidirectional composites. *Composites Part A: Applied Science and Manufacturing*, *127*, 105620.Sudhir, A. and Talreja, R., 2020. Corrigendum to "Simulation of manufacturing induced fiber clustering and matrix voids and their effect on transverse crack formation in unidirectional composites" [Compos. Part A: Appl. Sci. Manuf. 127 (2019) 105620]. *Composites Part A: Applied Science and Manufacturing*, *131*, p.105705.

In obtaining the results shown in Figure 8.44 circular voids were distributed randomly in the inter-fiber regions of the RVEs (Figure 8.28). To gain further insight into the local interaction between voids and fiber clusters another study [72] was conducted. In that study, RVEs of nonuniformly distributed fibers were first analyzed to locate the initiation of brittle cavitation as described above in Section 8.4.3.3. Assuming these sites to generate an initial debond of arc length subtending a small angle of 2° located at the fiber–matrix interface, the debond growth and kinking were analyzed. Figure 8.45 shows details of a typical case indicating the debond initiation site within a nonuniform fiber distribution pattern and a uniform fiber distribution pattern, the debond angle, and labeling of the two ends (tips) of a debond crack. Following previous studies [67, 68], the ERRs of the two debond crack tips were calculated for the nonuniform and uniform fiber distributions. Figure 8.46 shows the ERR variations with the debond angle for both cases. Comparing these results to those shown in Figure 8.35 it can be noted that the assumed debond initiation site in the latter results in symmetry in the debond growth with respect to the loading axis, while for the cases in Figure 8.46 the two crack tips have different ERRs and their different proximity to the neighboring fibers result in different influences on the ERRs of both crack tips. To understand how the presence of matrix voids affects the ERR variations it would suffice to consider one of the two debond crack tips.

Choosing the upper end of the debond crack a void was placed in three different positions shown in Figure 8.47 and in each case, three different void sizes were considered. In all cases, the voids showed an insignificant effect on altering

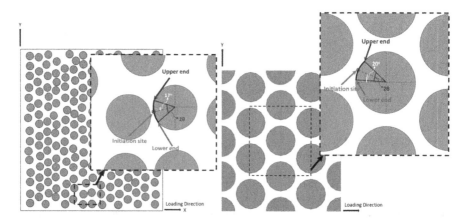

FIGURE 8.45 Details of the debond initiation site, the debond angle 2θ, and the surrounding of the debonded fiber within the RVE of 50% fiber volume fraction and 100% degree of non-uniformity (left) and within a uniform (hexagonal) fiber pattern (right). The initiation site in this case makes an angle of 17° clockwise with the debonded fiber's horizontal axis (loading axis) for nonuniform fiber distribution and 20° clockwise for uniform fiber distribution. Source: Elnekhaily, S.A. and Talreja, R., 2023. Effects of micro voids on the early stage of transverse crack formation in unidirectional composites. *Composites Part A: Applied Science and Manufacturing, 167,* 107457.

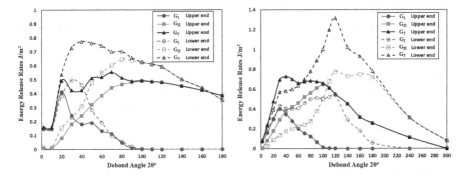

FIGURE 8.46 Variations with the debond angle 2θ of the ERRs of the two ends of a debond crack of initial debond angle $2\theta = 2°$ located within the nonuniform fiber distribution (left) and uniform fiber distribution (right). Source: Elnekhaily, S.A. and Talreja, R., 2023. Effects of micro voids on the early stage of transverse crack formation in unidirectional composites. *Composites Part A: Applied Science and Manufacturing, 167,* 107457.

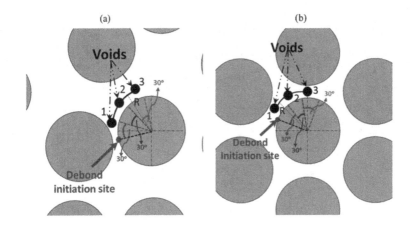

FIGURE 8.47 Placements of voids in positions marked as 1, 2, and 3, described by the radial distance R from the center of the debonded fiber and angles as shown for (a) non-uniform and (b) uniform fiber distributions. Source: Elnekhaily, S.A. and Talreja, R., 2023. Effects of micro voids on the early stage of transverse crack formation in unidirectional composites. *Composites Part A: Applied Science and Manufacturing, 167,* 107457.

the occurrence of brittle cavitation without voids. This suggests that the stress field induced by the voids does not have a strong dilatational component. Thus, if brittle cavitation is the cause of debonding, as has been assumed here, then the presence of microvoids will not be a significant factor. However, if interfacial normal stress is taken as the cause of debonding, as has been assumed, e.g., in [73], then a different conclusion regarding the effect of voids on debond initiation will result. It is our contention that the critical value of interfacial normal stress cannot be used for debond initiation as this value cannot be determined uniquely by a single-fiber experiment.

This is because, in a multiple-fiber pattern within a composite, the stress triaxiality at a fiber–matrix interface differs from that in a single-fiber case due to the mutual interaction of fibers. This fundamental difficulty also applies to cohesive zone models for interfacial failure because of the difference in mode mixity of decohesion in single-fiber versus multiple-fiber environments.

The effects of the size and location of voids on brittle cavitation have also been studied in the context of the constraining effect of longitudinal layers on transverse crack formation in a cross-ply laminate subjected to axial tension in [74]. In that study, large effects of microscopic voids on brittle cavitation have been found in regions of transverse plies close to the ply interfaces.

The effects of voids on the growth of debond cracks were studied in [72] by calculating the ERRs of the upper crack tip in the presence of voids of three sizes in three locations, as described in Figure 8.47 for both cases of nonuniform and uniform fiber distributions. Figure 8.48 shows the results for the nonuniform fiber distribution with

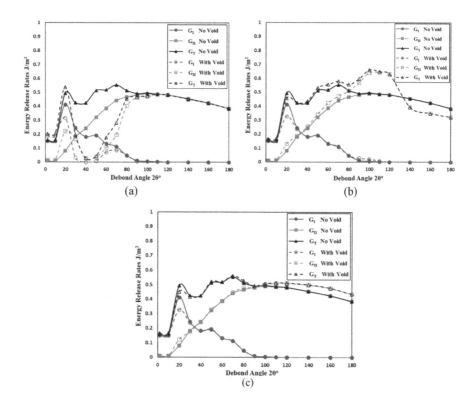

FIGURE 8.48 ERR variation along the upper end of the debond crack with the debond angle for a void of diameter 1.6 mm. (a), (b), and (c) correspond to void positions 1, 2, and 3, as shown in Figure 8.47(a). Source: Elnekhaily, S.A. and Talreja, R., 2023. Effects of micro voids on the early stage of transverse crack formation in unidirectional composites. *Composites Part A: Applied Science and Manufacturing, 167*, 107457.

the void of diameter 1.6 mm, and the results for uniform fiber distributions are shown in Figure 8.49. While the effects on the ERRs vary with void sizes and locations, it was found that these effects depend on the relative size and location of a void with respect to the surrounding fibers. The voids were found to be most effective in altering the ERRs when they were relatively away from fibers.

Finally, the kink out of a debond crack tip in the presence of a void was also studied in [72]. The position of the debond crack tip at which the kink out in the matrix occurs was determined by the maximum principal stress at the crack tip which was calculated at increasing debond angles. Comparing this position when a void is not present it was found that kinking occurs earlier when a void is relatively large and is close to the debond crack tip, as shown in Figure 8.50(b), while the kinking is delayed to a larger debond angles when the void is relatively small and is relatively away from the debond crack tip, as shown in Figure 8.50(c). The effect on

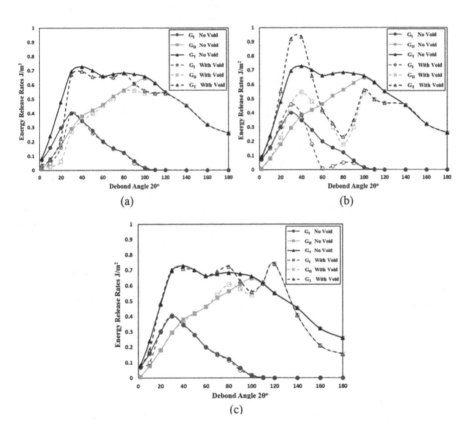

FIGURE 8.49 ERR variation along the upper end of the debond crack with the debond angle for a void of diameter 1.6 mm. (a), (b), and (c) correspond to void positions 1, 2, and 3, as shown in Figure 8.47(b). Source: Elnekhaily, S.A. and Talreja, R., 2023. Effects of micro voids on the early stage of transverse crack formation in unidirectional composites. *Composites Part A: Applied Science and Manufacturing*, *167*, 107457.

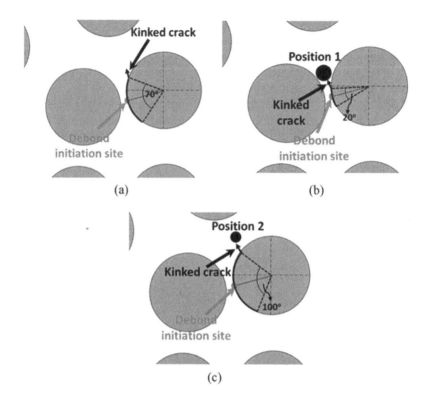

FIGURE 8.50 Positions of kinked debond cracks (upper ends) for (a) no voids, (b) a void of diameter 1.6 μm in position 1, and (c) a void of diameter 0.8 μm in position 2, all in a nonuniform fiber pattern. Source: Elnekhaily, S.A. and Talreja, R., 2023. Effects of micro voids on the early stage of transverse crack formation in unidirectional composites. *Composites Part A: Applied Science and Manufacturing, 167*, 107457.

when kinking occurs is dependent on the perturbation of the mode I ERR caused by the void as kinking occurs while the debond crack is still in the opening mode, as discussed above in Section 8.4.3.4.1. The effect of void on debond crack kinking in a uniform fiber pattern was also examined in [72]. Figure 8.51 illustrates one case where a void induces kinking earlier compared to the no-void case.

8.4.4 Transverse Compression

The observations related to the failure of UD composites subjected to transverse compression were reviewed in Chapter 5, Section 5.5. The early observations reported by [74] in 2007 were focused on the final failure plane and its orientation. The failure plane after fracture (separation) showed the presence of fiber–matrix debonding and hackles, which indicated the involvement of matrix shear in the failure process. The failure plane orientation was not at 45° suggesting that something more than pure shear was in play. Based on these observations the authors in [75] conducted a

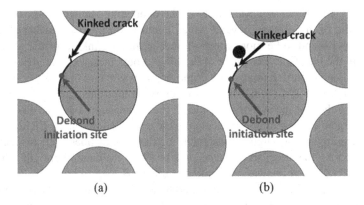

FIGURE 8.51 Kinking of a debond crack for (a) no void and (b) a void of diameter 0.8 µm. Note the reduced debond angle at kinking in the presence of a void. Source: Elnekhaily, S.A. and Talreja, R., 2023. Effects of micro voids on the early stage of transverse crack formation in unidirectional composites. *Composites Part A: Applied Science and Manufacturing, 167,* 107457.

computational micromechanics study. They assumed the polymeric matrix (epoxy) to behave like an elastoplastic solid whose plastic deformation is governed by the Mohr–Coulomb criterion, i.e., yielding initiates on a plane when the shear stress on the plane overcomes cohesion and frictional stress induced by the normal stress on the plane. The inclination of the plane then differs from 45° to the applied normal stress due to the (internal) friction modeled by the Mohr–Coulomb law. The authors accounted for fiber–matrix debonding by using a traction–separation interface failure model that requires interfacial "strength" and "fracture energy" as input. The normal and tangential strengths of the interface were assumed to be equal. It is noted at this point that these strength values cannot be determined experimentally as the stress state at the fiber–matrix is always triaxial and the triaxiality depends not only on the properties of fibers and matrix but also on the presence of neighboring fibers and the configuration in which they are placed. The interfacial fracture energy is also not a unique value as it depends on the quality of the bond, which in turn depends on the manufacturing process. The authors conducted a parametric study varying the strength and fracture energy as parameters. They found that the strength had the main influence on the failure while the fracture energy had a secondary effect. These conclusions should be viewed in the context of the assumptions made regarding the validity of the cohesive failure law and the equality of normal and tangential interfacial strength.

In 2008, the study [76] examined the failure plane orientation by focusing on the fiber–matrix debond crack assuming the kinking out of this crack into the matrix will govern the direction in which the final failure will occur. They considered a single fiber embedded in a polymeric matrix subjected to compression normal to the fiber. In a previous study [77], initiation of debonding in a perfectly bonded fiber was determined by the maximum interfacial shear criterion, and the propagation of an initial small debond crack was then analyzed by interfacial fracture mechanics concepts. Figure 8.52 displays details of the debonding showing the extent of the crack in the opening mode beyond which contact zones exist. Further crack progression was found to be more energy favorable in the kinking of the lower crack tip than in the interface. The angle of the kinked-out crack was then determined by the maximum circumferential stress criterion and was found to be $50° < \theta < 58°$ for glass–epoxy and $50° < \theta < 62°$ for carbon–epoxy composites.

In the study [78] the authors considered the influence of a second fiber near the fiber undergoing debonding. By calculating the perturbations on the ERR they found that depending on the angular placement of the second fiber with respect to the debonding fiber the effect was to accelerate or shield the debond growth. The effect on debond crack kinking was not studied in this work.

The two approaches described above addressing the assumed underlying mechanisms that govern the formation of an inclined failure plane under transverse compression are perhaps not complete explanations of the failure process. The approach in [75] emphasizes the role of matrix plasticity while also considering fiber–matrix interfacial failure by cohesive zones. The approach in [76] attributes the kinking of the interfacial debond crack as the governing cause of cracking along an inclined plane. The most detailed observations of failure under transverse compression [79] reported most recently suggest that more work is needed to provide a full explanation of the failure process. As described in Chapter 5, Section 5.5, these observations show the role of fiber–matrix debonding and kinking in the early stage of the failure process to be different than in the final stage of inclined plane formation. In the

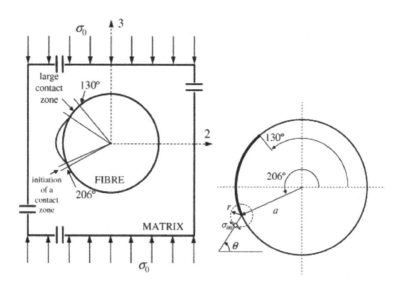

FIGURE 8.52 A single-fiber composite under transverse compression showing details of fiber–matrix debonding (left), and kinked-out debond crack (right). Source: Correa, E., Mantič, V. and París, F., 2008. A micromechanical view of inter-fibre failure of composite materials under compression transverse to the fibres. *Composites Science and Technology*, *68*(9), 2010–2021.

early stage, matrix cracks were found to form in the loading direction near the fiber–matrix interface, apparently not as kinked-out debond cracks. Toward the end of the failure process, the macrocracks in the matrix were on inclined planes. The role of matrix plasticity could not be inferred from the reported observations.

8.4.5 In-Plane Shear

As noted in Chapter 5, Section 5.5, testing under pure shear conditions is severely limited by practical problems arising from the necessity of introducing uniform shear stress in a large enough region to allow observing the failure process. At best what has been inferred from testing is that matrix failure in the inter-fiber region shows hackles on the fracture surface indicating plastic deformation. Prior to failure, multiple cracks in the matrix between fibers have been seen lying inclined to fibers. These cracks subsequently grow along fibers and join to form large cracks resulting in failure in the fiber direction. Thus, in-plane shear failure is described as an axial shear failure.

From experimental observations, it is not easy to infer which failure event initiated the failure process. It is likely that the inelastic deformation of the polymeric matrix is a precursor to the observed cracks. The computational micromechanics study [80] investigated the occurrence of yielding in UD composites under axial shear. A three-dimensional RVE was constructed following the algorithm developed in the previous work [55] with different fiber volume fractions and different degrees of fiber distribution nonuniformity. Finite element stress analysis was performed on RVEs using a thermal cooldown of 80°C followed by axial shear. Each RVE realization was searched to find the point in the matrix where the distortional energy density attained the maximum value. Figure 8.53 shows variations with the applied

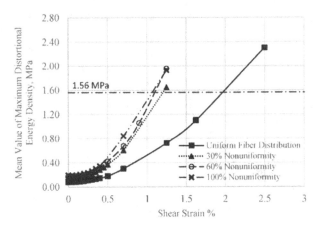

FIGURE 8.53   The maximum value of the distortional energy density averaged over multiple RVE realizations of different degrees of nonuniformity of fiber distributions, at 40% fiber volume fraction, plotted against the applied axial shear strain. The uniform fiber distribution case is also shown for comparison. Source: Elnekhaily, S.A. and Talreja, R., 2019. Effect of axial shear and transverse tension on early failure events in unidirectional polymer matrix composites. *Composites Part A: Applied Science and Manufacturing, 119,* 275–282.

axial strain of the maximum value of the distortional energy density averaged over multiple realizations of RVEs corresponding to 30%, 60%, and 100% degrees of nonuniformity at 40% fiber volume fraction of a glass–epoxy UD composite. The point in each RVE realization where the maximum distortional energy density was maximum in the matrix was found to be between the two closest-spaced fibers. This effect of fiber clustering is clearly seen in Figure 8.53, where compared to the uniform fiber distribution case all nonuniform fiber distributions attain the critical value of the distortional energy density (1.56 MPa) much earlier. This significant effect of fiber clustering was, however, found not to depend much on the degree of nonuniformity, as seen in Figure 8.53. The same trend was also found at higher fiber volume fractions. The occurrence of yielding was found to be affected significantly by the fiber volume fraction, as shown in Figure 8.54.

8.4.6 MODELING PLASTIC DEFORMATION OF A POLYMER MATRIX WITHIN A COMPOSITE

Modeling of failure beyond the first event consisting of initiation of yielding in a polymer matrix is subject to complexity because of the presence of fiber–matrix interfaces. As one scenario, these interfaces can be assumed to be perfectly bonded in which case plastic flow will occur in the polymer matrix under altered local stress state caused by the fibers. As a modification of this scenario, one can account for the

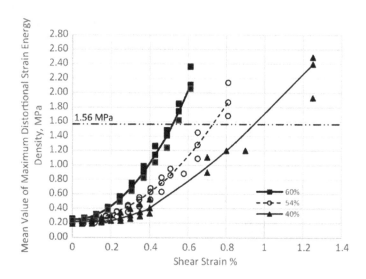

FIGURE 8.54 The maximum value of the distortional energy density attained in RVEs of different fiber volume fractions, averaged over multiple realizations, plotted against the applied axial shear strain. The degree of nonuniformity is 100% in all cases. Source: Elnekhaily, S.A. and Talreja, R., 2019. Effect of axial shear and transverse tension on early failure events in unidirectional polymer matrix composites. *Composites Part A: Applied Science and Manufacturing, 119*, 275–282.

structural changes in the polymer close to the fiber surface induced during processing. One study [81] used coarse-grained molecular dynamics (CG-MD) simulation to characterize the uniaxial mechanical properties of epoxies around fibers. These properties (peak stress and strain, and failure strain) were determined as a function of the distance from the fiber–matrix interface. A finite element model (FEM) was used to calculate local stresses in a composite unit cell consisting of a uniform fiber pattern subjected to transverse tension. This approach was described as structure-informed FEM. For comparison, the conventional FEM approach with the same epoxy properties everywhere without the influence of proximity to the fiber–matrix interface was used. A noteworthy difference in the prediction by the two approaches was that the structure-informed FEM predicted more localization of plastic deformation in the interface than in the traditional (uniform epoxy properties) case. Figure 8.55 [81] illustrates this difference. As seen, when the epoxy matrix is given uniform properties, the yielding is spread out from the fiber–matrix interface, while with varying properties the yielding in the epoxy matrix is localized at the interface. Although the criterion applied in these computations is for "ductile damage", the progression of the damage has features commonly seen in transverse cracking, as discussed above.

While study [81] brought out the effect of structural changes (degree of cross-linking and free volume size) in epoxies on the deformation and ductile failure in

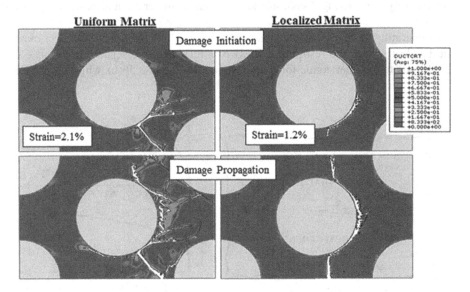

FIGURE 8.55 Initiation and progression of yielding in the epoxy matrix of a composite loaded in transverse (horizontal) tension. The "uniform matrix" refers to the case of the same epoxy properties irrespective of the distance from the fiber surface, while "localized matrix" is the case where properties are dependent on the distance from the fiber surface. Source: Wu, X., Aramoon, A. and El-Awady, J.A., 2020. Hierarchical multiscale approach for modeling the deformation and failure of epoxy-based polymer matrix composites. *The Journal of Physical Chemistry B*, *124*(52), 11928–11938.

composites, it did not consider the role of stress triaxiality on brittle cavitation in epoxies. The work described above in Section 8.4.3.1 clearly establishes this role. Of note is the MD simulation study [31] that described the formation of brittle cavities under hydrostatic tension. Since these cavities form close to the fiber surface, they are likely to initiate debonding at the interface, thereby diverting further progression of failure to debond growth and kinking, as discussed above. Therefore, ignoring this and focusing instead on matrix plasticity for transverse crack formation is questionable.

Brittle cavitation in epoxies is, however, not always a favorable failure event. It only occurs if the dilatational energy density at a point within the epoxy volume reaches its critical value for cavitation. As noted in [80], under axial shear imposed on a UD composite, yielding instead of brittle cavitation occurs. Further progression of failure events is then inelastic deformation leading to ductile failure.

8.4.7 MODELING INTERFACE DEBONDING BY COHESIVE ZONES

The initial concept of cohesive zones is credited to the works of Dugdale [82] and Barenblatt [83]. In his classical paper on the J-integral, Rice [84] explained the Dugdale–Barenblatt crack model as an alternative to the sharp-tipped crack model where a singular stress field arises for linear elastic materials. The singularity is removed by considering a cohesive zone ahead of the crack and postulating that the influence of atomic attraction can be represented by a restraining stress on the separating surfaces, as schematically described in Figure 8.56 [84]. For elastic brittle fracture, the restraining stress σ is related to the separation distance δ analogous to the interatomic force-separation law, Figure 8.56(b), while for ductile

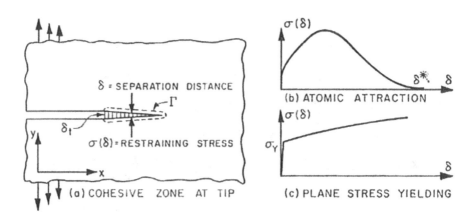

FIGURE 8.56 Dugdale–Barenblatt crack model, (a) cohesive zone at the crack lip with restraining stress dependent on separation distance, (b) force-displacement relation for atomic attraction in elastic brittle fracture, and; (c) for plane-stress plastic yielding in a thin sheet. Source: Rice, J.R., 1968. A path independent integral and the approximate analysis of strain concentration by notches and cracks. *Journal of Applied Mechanics*, Trans. ASME, 379–386

fracture due to plane stress yielding in thin sheets, the stress–separation is as shown in Figure 8.56(c). The separation distance δ^* for elastic brittle fracture is when the atoms at the crack tip can be pulled out of the range of their neighbors. It is worth noting that the σ–δ relationship and δ^* are conceptual entities that are generally not possible to determine by first principles. However, Rice showed that for small dimensions of the cohesive zone ("small-scale yielding") the σ–δ relationship and δ^* were immaterial as the fracture theories with or without a cohesive zone were identical.

The use of cohesive zones at a crack tip, as described above (Figure 8.56), was extended to the case of interface failure without a pre-existing crack (debond) in [85]. In this study, the authors analyzed the initiation of debonding at the surface of a rigid particle embedded in a crystal matrix. Assuming normal and shear traction–separation relationships for the interface, the authors used a critical separation at the interface as a criterion for debond initiation. This use of CZM without a pre-existing crack was novel and led to many other studies of interface failures.

Among the first studies to employ CZM for interface debonding in polymer matrix composites was [86]. Describing their CZM as a cohesive crack model, the authors assumed the interface decohesion to follow the traction–separation relationship shown schematically in Figure 8.57 where the normal traction across the interface is plotted against a generalized crack surface separation measure given by λ. This measure is constructed as an assumed function of the three relative displacements

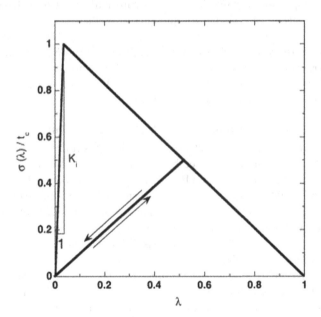

FIGURE 8.57 A schematic description of the interface decohesion model. The stress s is the normal stress transferred through the cohesive crack whose separation is described by a parameter λ. The parameter τ_c is the interfacial strength. Source: Segurado, J. and LLorca, J., 2005. A computational micromechanics study of the effect of interface decohesion on the mechanical behavior of composites. *Acta Materialia*, *53*(18), 4931–4942.

of the crack surfaces (interfacial separations). The initial interfacial "stiffness" K_i is an adjustable value.

The model described above is typical of the models presented in the literature as interface "cohesive zone models" (CZMs). Such models have inherent uncertainties because the parameters described there as "strength" and "toughness" are strictly not material properties. The values of these parameters are calibrated by adjusting the calculated outcomes to selected reference results. One study [87] used such estimates of the parameters and varied their values to see how that affected the RVE-averaged stress–strain response and transverse failure of a carbon–epoxy UD composite. While such parametric studies are useful in gaining insight into the nature of the fiber–matrix interface, the quantitative predictions such as the transverse failure strain by such models cannot be relied upon. In fact, a wide variety of assumptions are possible to construct CZMs and conduct parametric studies using those. One example to illustrate this is a study [88] that assumes a trapezoidal CZM where the traction-separation relationship is given a constant peak value over a range $\lambda_1 \leq \lambda \leq \lambda_2$ rather than at one λ value as in Figure 8.57, providing thereby one additional adjustable parameter. This study also examined the effects of providing two different strengths to the interface. They used the standard isotropic J_2-flow theory and did not consider the thermal cooldown of the composite. Suffice it to say that numerous such parametric studies are possible, and each study can at most be assessed by a qualitative comparison with trends in the experimentally observed behavior. Caution is warranted in using such studies for quantitative predictions.

The proliferation of CZMs in recent years is driven by their incorporation into commercial software [89]. This easy access to CZMs in FE stress analysis is an attractive and convenient feature but has the risk of improper use if the physical basis for CZMs is not kept in sight. The following discussion is meant to clarify the underlying physical considerations for CZMs.

8.4.7.1   Fiber–Matrix Debonding as a Multiscale Problem

If the fiber–matrix bond is "perfect", i.e., the interface is fully bonded with no gaps, then the debonding can be addressed by assuming a reasonable process of debonding. If, on the other hand, the bonding is not perfect, then specific knowledge of imperfections would be required for that assessment. Due to a range of possible processes by which composite materials are manufactured, it is difficult to know even in the perfect bonding case how the fiber–matrix interfaces are generally formed. For polymer matrix composites such as glass–epoxy and carbon–epoxy, all possible processes of forming interfaces (van der Waal forces, acid–base interactions, and chemical bonds) can for simplicity be summarized for interfacial characterization in terms of surface concentration of interfacial bonds and the bond energies [90]. Thus, two quantities, "strength" and "energy", together would be needed to characterize an interface.

The problem of determining strength (understood as a peak value of stress at the interface) and energy expended in causing interfacial failure resulting in a debond (crack) is not straightforward. Figure 8.58 depicts the three-level problem [90], viz., at the molecular level, Figure 8.58(a), at the single-fiber level, Figure 8.58(b), and at

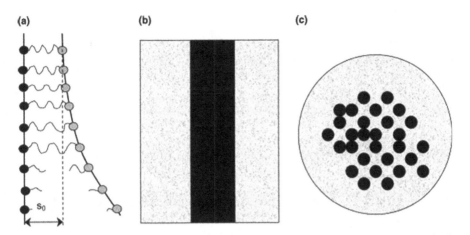

FIGURE 8.58 Characterization of fiber–matrix debonding described as a three-level problem: (a) Molecular level, (b) single-fiber composite level, and (c) composite level. The distance s_0 at the molecular level is the equilibrium interatomic distance. Source: Zhandarov, S. and Mäder, E., 2005. Characterization of fiber/matrix interface strength: applicability of different tests, approaches and parameters. *Composites Science and Technology*, 65(1), 149–160.

the composite (multifiber) level, Figure 8.58(c). From the composite failure analysis point of view, it is important to note that the fiber distribution within a composite is nonuniform, as depicted in Figure 8.58(c). This, combined with properties of fibers and matrix (bulk and in-situ) will result in different stress states at fiber–matrix interfaces and their variations around the fibers under given loading modes (axial tension–compression, transverse tension–compression, in-plane shear, and combinations of these). These stress states are generally triaxial and they vary in their triaxiality ratios (e.g., expressed in terms of ratios of the largest principal stresses to the mean stresses). To characterize failure caused at the interface by a peak value of a stress component (or principal stress) will be incorrect unless the stress state is hydrostatic tension, i.e., with one stress component only, as done in [54] initially, and continued later in [55], [56] and [72], as described in Section 8.4.3 above for the case of transverse tension. In fact, in these works, no cohesive zone is assumed at fiber–matrix interfaces and failure is assumed to occur in the matrix. Several works that use cohesive zones have used two separate traction–separation laws, one in tension and the other in shear, followed by an "interaction" criterion. There are many examples of this, e.g., [91–93]. Figure 8.59 describes a typical modeling approach proposed in [92] and used in [93].

There is no fundamental reason for the interaction (combined effect) to be described by a quadratic relationship, but it is commonly done as a convenient way to deal with an intractable problem. Once the initiation of failure at a point on the fiber-matrix interface is determined by using a preselected "strength" of the interface, the debond crack is allowed to form by using the area under the bilinear traction-separation law (see Figure 8.59, right), i.e., the fracture energy, which provides the extent of the cohesive zone. It is noted that the procedure proposed in [92] and depicted in

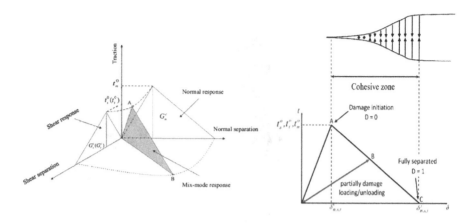

FIGURE 8.59 Two separate bilinear traction–separation laws for normal and shear tractions combined with a quadratic interaction relationship. Sources: Zhang, C., Li, N., Wang, W., Binienda, W.K. and Fang, H., 2015. Progressive damage simulation of triaxially braided composite using a 3D meso-scale finite element model. *Composite Structures*, *125*, pp.104-116, and Palizvan, M., Tahaye Abadi, M. and Sadr, M.H., 2020. Micromechanical damage behavior of fiber-reinforced composites under transverse loading including fiber-matrix debonding and matrix cracks. *International Journal of Fracture*, *226*, 145–160.

Figure 8.59 is a variation of what was proposed in [86] and depicted in Figure 8.57. The fracture energy is assumed to have two basic modes in opening and sliding separations in the cohesive zone, analogous to interfacial cracks. The combined effect is assumed to follow some mixed-mode criterion used for regular cracks, ranging from linear to nonlinear to combinations of the two. For nonlinear criteria, the choice is often quadratic unless an additional adjustable parameter is needed in which case another form, e.g., a polynomial is used.

As noted above, the use of cohesive zones in treating interface failure – both initiation and progression of debonding – has accelerated in recent years driven by the convenience of incorporating these in a finite element computation scheme. The justification offered with respect to their physical basis is often a perceived failure process at the interface schematically depicted at the molecular level in Figure 8.57(a). It is important to recognize that no measurements made at the molecular level are, however, involved in estimating the cohesive zone modeling parameters. These parameters need to be quantified at the single-fiber level, Figure 8.57(b), as a manifestation of the molecular failure behavior in the solid mechanics framework. One study [94] examined the cohesive zone models proposed in the literature and found a wide range of values of cohesive zone parameters (critical separation length and peak stress) that were assumed. The critical separation length in the models varied from 10^{-10} to 10^{-5} m, and the peak stress (often called "strength") varied from MPa to GPa. The "fracture toughness", i.e., the area under the force–separation curve (e.g., bilinear, or exponential) also varied accordingly. The inferences drawn from using different laws were also not the same. The authors compared the load–displacement curves obtained in a fiber push-out test for a metal matrix composite using an exponential traction–separation law and a bilinear law and found significant differences in the predictions by the two models. By examining the sensitivity of the model parameters to model predictions, they concluded that for physically meaningful results all parameters need to be determined accurately including the shape of the traction-separation relationship.

The composite level prediction of fiber-matrix debonding requires calculating the stress states at and near the fiber surfaces. Since the stress triaxiality was shown to be important in initiating debonding [54–56], the inclusion of thermal cooldown-induced residual stresses in polymer matrix composites is necessary for the accurate determination of the dilatational energy density. Furthermore, as described in Section 8.4.3.3 above, the nonuniformity of fiber distribution has a significant impact on the fiber–matrix debond initiation. Therefore, a measure of this nonuniformity should also be associated with the computation of inter-fiber stress states under given external loading. With respect to cohesive zones, it is worth pointing out that an *a priori* insertion of cohesive zones at fiber-matrix interfaces is likely to alter the local stress triaxiality. If so, the effect of stress triaxiality on debond initiation will not be assessed properly.

At this point, it is also worth noting that the number of independent CZM parameters in a general case is seven: Two cohesive strengths (tension and shear), two fracture energies (separation and sliding), two critical separations (normal and shear), and one mode-mixity adjustment parameter (depending on the curve-fitting scheme). To evaluate these parameters by independent tests is likely not possible. To proceed with CZM analyses simplifications are often introduced, e.g., equal decohesion strengths in tension and shear (physically unreasonable), or a quadratic relationship connecting the two strengths. Similarly, fracture energies are either assumed equal (also physically unreasonable), or a mode-mixity adjustable parameter is used based on LEFM. Finally, the critical separation in the traction-separation relationship results from assumed relationships (often called "shape" of the traction-displacement "law"). In view of the inability to determine the CZM parameters by independent tests, the use of CZM leads to uncertainties in results. It would be reasonable, therefore, to exercise caution in drawing conclusions based on CZM-based failure analyses.

The study [94] examined a wide range of proposed CZMs to check their consistency in the thermodynamic sense, i.e., proper energy dissipation in debonding at interfaces. Only one traction–displacement relationship was found to be consistent when the exponential traction–displacement originally proposed in [85] was modified to allow unequal work of separation in normal and shear modes. That model was further modified by introducing an (unspecified) internal damage variable to derive the traction–displacement relations from a potential. This additional correction indicates the difficulties of using CZM reliably despite the convenience of the numerical implementation of these models.

8.4.7.2   Fiber–Matrix Debond Growth and Kink Out

The debond crack growth along the fiber–matrix interface and subsequent kink out into the matrix using the linear elastic fracture mechanics (LEFM) approach was described above in Section 8.4.3 where the effects of fiber distribution non-uniformity and matrix voids were also discussed. The same problem has also been addressed by cohesive zone modeling (CZM). It would be useful to examine the two modeling approaches to understand their relative advantages and disadvantages.

The study [95] applied CZM to debond crack growth for three cases: a) A single fiber, b) two fibers, and c) multiple fibers uniformly or nonuniformly distributed. Cohesive zones were placed all along the fiber circumference and the composite was loaded in transverse tension. Results for the single-fiber case showed debond growth as two circular arcs growing symmetrically about the loading axis. This debond crack shape and symmetry were also found for two fibers placed along the loading axis, across the loading axis, or at 45° to the loading axis. The multifiber case of uniformly distributed fibers showed unrealistic simultaneous growth of debonding on all fibers. The real-life situation was modeled with randomly distributed fibers, in which case the debond growth was found to be random partly due to the random initiation of debonding assumed by a strength criterion and partly because of the randomness of the driving force for decohesion. Figure 8.60 shows results for two different fiber radius values and two values of remotely applied transverse tension. The debonding shows no indication of the commonly observed transverse cracks.

The study [96], published before [95], reported the CZM results for the single, double, and multiple fiber cases with the aim of comparing the CZM approach with the LEFM approach. This study used the bilinear traction–displacement relationship for decohesion assuming the same values for normal and shear responses. For decohesion initiation, the authors took the maximum traction in the traction–displacement relationship as the same stress value for unstable debond growth in a fiber fragmentation test reported in [97]. It is noted that this stress value is for unstable growth of a pre-existing crack in mode I and cannot be taken to indicate "initiation" of a debond crack. Clearly, an unstably growing crack does not have a second linear traction–displacement line in the bilinear relationship of the type displayed in Figure 8.57. Furthermore, the unstable debond crack growth in pure mode II will be different from this value, and assuming the two values to be the same will introduce errors in analyzing debond initiation that likely cannot be assessed. The fracture energy (area under the traction–displacement triangle) was also taken to be the same as LEFM-defined critical ERR. Therefore, a proper basis for comparison of CZM with LEFM cannot be relied upon.

The kink out of a debond crack into the matrix has been treated by LEFM [67], as described above in Section 8.4.3.4.1. Much before the results described there were obtained a study [98] attempted to analyze the kink-out problem using cohesive zones. The model parameters used were those obtained by LEFM, e.g., the critical ERRs, and the kink out was assumed to occur when the ratio of G/G_c of crack advancing in the matrix exceeded the same ratio for the advancement of debonding. This is essentially an LEFM approach and using cohesive zones does not seem to offer any advantage.

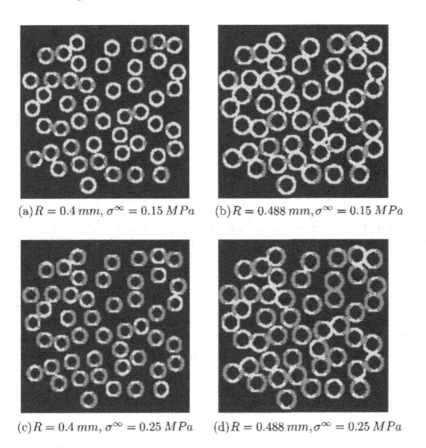

(a) $R = 0.4\,mm$, $\sigma^\infty = 0.15\,MPa$ (b) $R = 0.488\,mm$, $\sigma^\infty = 0.15\,MPa$

(c) $R = 0.4\,mm$, $\sigma^\infty = 0.25\,MPa$ (d) $R = 0.488\,mm$, $\sigma^\infty = 0.25\,MPa$

FIGURE 8.60 Debond crack growth determined by cohesive zone modeling for fibers of two different radius R distributed randomly in the UD composite cross section and loaded in transverse tension. Source: Dimitri, R., Trullo, M., De Lorenzis, L. and Zavarise, G., 2015. Coupled cohesive zone models for mixed-mode fracture: A comparative study. *Engineering Fracture Mechanics*, *148*, 145–179.

8.4.8 MODELING INTERFACE DEBONDING BY PHASE FIELD CONCEPTS

Another approach to fracture problems that has received increasing interest is based on the phase field concept. This concept originated in the desire to treat an interface not as a sharp transition from one material phase to another but more as a "diffuse" interface, often called an "interphase" [99]. A phase field variable is introduced as an independent thermodynamic state variable that tracks the evolution of the microstructure in the phase toward equilibrium. The utility of the state variable was later found in describing brittle fracture in the interphase by including it in the energy balance consideration originally formulated in the Griffith theory, see [100, 101] for review. The concept is illustrated in Figure 8.61 where the state variable ϕ is given the value $\phi = 0$ for a finite-sized intact diffuse interface and $\phi = 1$ stands for the

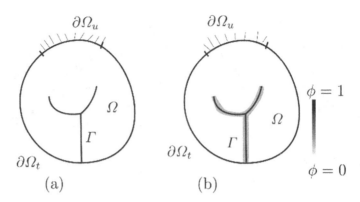

FIGURE 8.61 Conventional sharp cracks (a) replaced by "crack bands" and (b) with a phase field variable φ to account for the interface "band". Source: Wu, J.Y., Nguyen, V.P., Nguyen, C.T., Sutula, D., Sinaie, S. and Bordas, S.P., 2020. Phase-field modeling of fracture. *Advances in Applied Mechanics*, 53, 1–183

cracked state. The width of the interface "band" is incorporated as a material length-scale parameter in the fracture toughness value (critical ERR).

The conventional Griffith criterion for elastic brittle fracture, i.e., initiation of growth of a crack of surface area A by an incremental size dA can be expressed as,

$$\frac{d\Pi}{dA} = \frac{d\psi}{dA} + \frac{dW}{dA} = 0 \qquad (8.13)$$

where Π is the total energy, ψ is the elastic strain energy density, and W is the work needed to create new surfaces in advancing the crack. The last term is also called the critical ERR or fracture toughness, G_c. Thus, the pre-existing crack will grow if the strain energy stored in the volume exceeds the fracture toughness. A variational form of the criterion is written as,

$$\Pi = \int_\Omega \psi \, dV + \int_\Gamma G_c \, d\Gamma \qquad (8.14)$$

where Ω is the domain volume containing the crack of total surface area Γ (Figure 8.61). The criterion for the crack advance is now expressed as a minimization of the total energy Π [101]. In phase field modeling the total energy is approximated by the following volume integral.

$$\Pi_l = \int_\Omega \left[g(\phi) \psi_0 + G_c \gamma(\phi, l) \right] dV \qquad (8.15)$$

where ψ_0 is the strain energy density of the uncracked solid and functions g and γ are introduced to modify the strain energy density and fracture toughness appearing in the Griffith criterion (8.14). Note that the length scale is a variable in the function γ that is introduced to modify the fracture toughness. The use of the phase field variable ϕ is justified here as an evolutionary variable to represent material degradation (damage), $0 < \phi < 1$, and the length-scale parameter l is interpreted as a regularizing parameter representing the finite crack band width that approaches zero when the band in the limit becomes a crack (e.g., an interface debond).

Several problems of brittle cracking have been treated with the phase field modeling approach and the solutions have been found to agree well with the classical (conventional) Griffith theory predictions [101]. However, the agreement has been found to depend on the assumed functions g and γ appearing in Equation (8.15). The main advantage of resorting to the phase field fracture approach is claimed to be the ability to track the crack path without introducing additional failure criteria related to the debond crack kink-out angle, as done in [65, 67]. However, this advantage should be weighed against the large number of modeling parameters in the phase field approach that must be calibrated by simulations.

8.4.9 INCORPORATING MANUFACTURING DEFECTS IN THE COHESIVE ZONE AND PHASE FIELD MODELS

The manufacturing defects in UD composites are mainly fiber distribution nonuniformity, matrix voids, and defective (imperfect) fiber–matrix interfaces. The effects of fiber distribution nonuniformity and matrix voids were treated above with and without invoking cohesive zones and phase field concepts. Incorporating defective interfaces in failure analysis requires knowing the nature and extent of the interfacial defects. If such knowledge is available, then these defects can be treated as preexisting cracks at the interfaces in conducting the LEFM-based analysis of debond crack growth and kink out described above. The question now is: What would be the advantage of resorting to cohesive zone and phase field fracture methods in comparison to the LEFM-based methods? In addressing this question, let us consider the failure analysis in three stages: 1) The initiation of the first failure event, 2) the progression of the next and subsequent failure events, and 3) the final failure.

8.4.9.1 Initiation of the First Failure Event

Assuming no cracks are present, the first failure event can be the initiation of a crack. For the first crack to form, two possibilities can be conceived: 1) A weak plane exists that has the potential to break open as a crack and 2) the deformation energy is concentrated at a point causing intense deformation leading eventually to crack formation by interatomic bond breakage. The first possibility has been addressed with the cohesive zone methods by placing potential planes of crack formation at the fiber–matrix interfaces, e.g., in [93, 95, 98]. As described in Section 8.4.7.1 (see Figure 8.57), the initiation of debonding is assumed when the stress in the cohesive zone reaches the peak stress in the traction–separation relationship. In an RVE consisting of nonuniform (random) fiber distribution, this condition for debond initiation will be satisfied at one or few cohesive zones among all assumed cohesive zones at fiber–matrix interfaces. For the second possibility to occur, the matrix in the inter-fiber region must begin the process of crack formation by reaching first a yield condition. In [93], the epoxy matrix was assumed to yield following the Drucker–Prager yield criterion. Thus, both debond initiation at fiber–matrix interfaces and yield initiation in the matrix were assumed by stress criteria in which critical stresses are regarded as material constants. While the matrix properties can be determined by independent tests, it is not clear how the critical tractions at the fiber–matrix interface can be found. Clearly, in any loading that results in failure at the interface, the stress state at failure of the interface will be triaxial. Thus, separating tensile failure stress and shear failure stress at an interface is not possible. It is common, therefore, to estimate the tensile failure stress at an interface by a test where the normal radial stress at the interface dominates compared to the shear stress. Similarly, the shear strength is estimated by a test where the normal radial stress is small compared to the shear stress. Then, a combined stress criterion, e.g., by using a quadratic polynomial function is assumed. Obviously, there is no physical basis for such a criterion, and this type of arbitrariness is likely to induce uncertainty in the failure analysis. The tensile and shear strengths found by different methods produce different values. For instance, in [93] the tensile and shear strengths used were 50 MPa and 75 MPa, respectively, for carbon–epoxy composites, while in [102] the peak shear stress in a bilinear traction-separation cohesive relationship of carbon–epoxy was reported to be approximately 25 MPa. In [103] for an S-glass–epoxy interface, the same stress was found to be 120 MPa [103].

It is obvious that an approach to failure that uses a material failure value by an independent test should be preferred to any approach that calibrates the modeling parameters from simulations using the model itself. In [55], [56], and [80] no cohesive zones were assumed at fiber–matrix interfaces. Instead, failure at such interfaces was assumed to occur when unstable cavity growth occurs in the matrix close to a fiber–matrix interface. Following the previous works [28–31], the unstable expansion of a cavity formed in the epoxy matrix, described as brittle cavitation, was characterized by the critical value of the dilatational strain energy density, which was found by an independent test on epoxy in the bulk form. For the initiation of the competing mechanism of matrix yielding, the critical value of the distortional strain energy density, also found by an independent test on bulk epoxy was used. The use of strain energy density to characterize failure initiation at a point in both failure modes accounts for stress triaxiality in a consistent manner.

The strain energy density approach for initiation of the first failure event in the presence of matrix voids surrounded by nonuniform fiber patterns has also been demonstrated in [56], [72], and [80] and described above in Section 8.4.3.5.

8.4.9.2   Progression of Failure after the First Failure Event

When failure at the fiber–matrix interface initiates, it is reasonable to assume that a small debond crack is formed. For a circular fiber, the arc length of such a debond crack can be taken to be a small fraction of the fiber circumference. Further progression of the debond crack is assumed in cohesive zone approaches to be decided by the decohesion process represented by the assumed traction–displacement relationship, as described above in Section 8.4.7.2. As pointed out there, a total of seven unknown model parameters are needed to complete the debond growth analysis.

Even after that, only the progression of debonding along the fiber–matrix interface can be analyzed since the cohesive zones placed along the interface guide the crack growth path. To address the experimentally observed debond kink-out the cohesive zone approach is combined with the phase field fracture approach as the latter has the capability to predict a crack path. However, two additional modeling parameters must be assessed that appear in Equation (8.15). Thus, a total of nine parameters (seven in cohesive zones + two in the phase field) are required to complete the crack kink-out analysis. Further growth of the kinked-out crack depends on the conditions such as the crack growth resistance offered by the matrix material. It should be realized that the stress field in the matrix is affected by the presence of the assumed cohesive zones and/or crack band assumed in the phase field model. Such zones and bands tend to stiffen the material near the fiber–matrix interface resulting in erroneous calculation of the stresses in the matrix. Therefore, the crack path after the debond kink-out predicted by a phase field model is likely to be questionable.

8.4.9.3   Final Failure

As is widely known, a UD composite under tensile loading inclined to the fiber direction fails by the unstable growth of a crack. While the early stages of the failure process are affected by microstructural defects such as nonuniformity of fiber distribution in the cross section and matrix voids, resulting in micro cracks at different locations in the cross section, the final failure occurs from the growth of the crack whose energy release rate reaches the fracture toughness first. Thus, the crack tip stress field dominates the final failure. Using cohesive zones at fiber–matrix interfaces will have little benefit when a crack has grown beyond a few fiber diameters in size. If, however, the phase field fracture approach is combined with cohesive zone modeling, the path of a growing crack can be predicted and the instability of its growth leading to final failure can in principle also be predicted. This approach will require knowing the crack growth resistance, which is not the bulk value of the matrix fracture toughness as the presence of fibers ahead of the growing crack tip will influence this value. Thus, this will add one more unknown to the set of unknown modeling parameters in the combined cohesive zone and phase field approach.

   To illustrate the inadequacy of cohesive zone modeling in predicting the final stage of failure in a UD composite, let us consider a study that examined the effect of matrix voids on the failure of a UD composite under transverse tension and compression [104]. This study distributed voids in the inter-fiber regions and in the matrix with an overall void volume fraction as a parameter ranging from 1 to 5% within a composite of nonuniform fiber distribution. The composite was simulated by randomly distributing 70 fibers in a RVE resulting in a fiber volume fraction of 60%. The glass fibers were taken as isotropic linear elastic while the epoxy matrix was modeled as a Drucker–Prager material. Cohesive zones placed at the fiber-matrix interfaces were given a tri-linear traction-separation behavior with the second branch of the relationship being horizontal representing constant-stress yielding (see traction–separation relationships described in Section 8.4.7 above).

   Figure 8.62 shows the numerically calculated stress-strain plot for the transverse tension loading response. As indicated, the predicted fiber–matrix debonding is

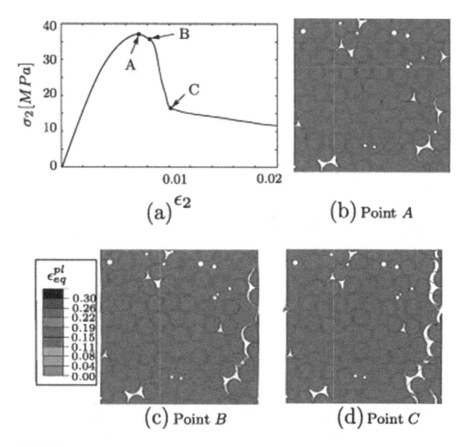

FIGURE 8.62 Overall transverse tensile stress-strain response of an RVE containing 70 randomly distributed glass fibers (fiber volume fraction 60%) and voids of 2% volume fraction. The images shown corresponding to points A, B, and C on the stress–strain plot display contours of equivalent plastic strain and fiber–matrix debonding. Source: Vajari, D.A., González, C., Llorca, J. and Legarth, B.N., 2014. A numerical study of the influence of microvoids in the transverse mechanical response of unidirectional composites. *Composites Science and Technology*, 97, 46–54.

dispersed at the peak stress and only results in a significant transverse crack past the peak stress and at the end of abrupt softening. This behavior does not correspond to experimentally observed stress–strain response which shows abrupt failure at the peak stress. As examples, experimental stress–strain responses of glass–epoxy composites of approximately 50% fiber volume fraction are shown in Figure 8.63 [105]. The glass fibers were either untreated (uncoated) or given sizing treatments of different types. Label A in Figure 8.63 refers to commercial sizing, while label B is for an elastomer coating, and labels C and D are for two sizings with high epoxy functionality. As seen in the figure, the high functionality sizings improve the transverse strength and strain to failure, while the uncoated, elastomer-coated, and commercially treated fibers provide approximately the same composite strength.

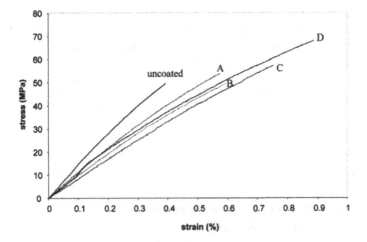

FIGURE 8.63 Transverse stress–strain response of a UD glass–epoxy composite in conditions where the fibers are uncoated and given different sizing treatments labeled as A, B, C, and D. Source: Benzarti, K., Cangémi, L. and Dal Maso, F., 2001. Transverse properties of unidirectional glass/epoxy composites: influence of fibre surface treatments. *Composites Part A: Applied Science and Manufacturing, 32*(2), 197–206

The significance of effective sizing C and D is to improve the fiber–matrix bonding and thereby delay debonding, resulting in a higher contribution of epoxy deformation (ductility) in the failure process. Fracture surfaces showed residues of the failed epoxy on the fiber surfaces, while cleaner fibers were seen for the uncoated case. In all cases, failure was found to result from interconnected debond cracks [105]. The lack of correspondence of the predicted stress–strain response (Figure 8.62) to the experimental response (Figure 8.63) is likely because in the failure analysis in [104] the debond crack kink out to connect with the neighboring debond cracks was not considered. The failure of the matrix as a ligament between debonded fibers was instead the final failure mode leading to the predicted softening.

8.5 FAILURE MODELING OF LAMINATES

Failure mechanisms in laminates in the presence of defects were reviewed in Chapter 6. Understanding these mechanisms even without defects took many years of research. Therefore, in the early years of composite design, it was common to focus on the "first ply failure", often referring to the initiation of transverse cracking within a ply of a laminate. Later, however, it was realized that a composite laminate can perform safely much beyond this failure event, and modeling the progression of failure events culminating in the final failure of a laminate was therefore deemed important.

Modeling of the first failure event and progression of the subsequent failure events to the criticality of final failure was discussed above for UD composites under different loading conditions. When a UD composite acts as a ply within a laminate, the constraint imposed by the other plies causes changes to its failure behavior. For instance, an unconstrained UD composite fails by unstable growth of a single transverse crack, but the same composite placed as a transverse ply in a cross-ply laminate gets that crack arrested by the longitudinal plies and the subsequent failure events then take the form of multiple cracking within the ply, as described in Chapter 6. Also, as described in that chapter, the transverse crack fronts induce local delamination at the interfaces of the longitudinal and transverse plies. Thus, beyond the failure mechanisms treated above for UD composites, modeling of failure in a cross-ply laminate involves considerations of multiple transverse cracking and local (also called interior) delamination. The same holds for general laminates where additionally these mechanisms must be considered for plies of multiple orientations.

8.5.1 FIRST FAILURE EVENT AND EARLY STAGES OF FAILURE IN CROSS-PLY LAMINATES

We begin first by examining how the two common manufacturing defects, namely, nonuniform fiber distribution and matrix voids, induce a first failure event in a UD composite that is constrained by the adjacent plies in a laminate. An effective way to do this is by studying crack formation in the transverse ply of a cross-ply laminate. A recent study [74] used computational micromechanics for this purpose. Following the failure analysis aimed at determining the first failure event in a UD composite (unconstrained ply), described above in Section 4.3.3 [55, 56], the study computed the local dilatational and distortional energy densities at matrix points within RVEs containing randomly distributed fibers and voids when the UD composite was placed as a 90° ply in a cross-ply laminate. Figure 8.64 displays the computational scheme where the outer 0° plies are homogenized while the inner 90° ply has randomly distributed fibers. The fiber clusters and resin-rich regions in the 90° ply were identified and voids were placed in different specified positions such as the resin-rich regions. The peak values of distortional energy density (U_d) and the dilatational energy density (U_v) are compared in Figure 8.65 for the unconstrained and constrained plies.

Several interesting features are evident in Figure 8.65. As reported earlier in [55, 56], the study [74] also found that the criterion for brittle cavitation, i.e., U_v reaching its critical value, is satisfied before yielding epoxy in the unconstrained UD composite. The U_v value was not affected much by the constraint on the ply (black and green colors in Figure 8.65), but that changed when a void of 3 mm diameter was placed in different locations within the RVEs. The enhancement in U_v was most when a void was placed near the fiber–matrix interface in the unconstrained ply. However, constraining the ply reduced this U_v value but not affected the void-induced enhancement at other locations. The study [74] also found a significant effect of void size on the initiation of brittle cavitation.

The effects of voids on the enhancement of U_d are similar, as Figure 8.65 shows, but the significance of this will be on the yielding of epoxy after brittle cavitation has occurred. The debonding at the fiber–matrix interface caused by brittle cavitation will likely alter the stress states in the epoxy matrix affecting its yielding and post-yielding deformation.

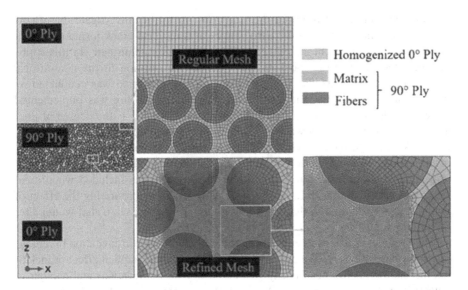

FIGURE 8.64 Details of the finite element model for computing the local stress field in a cross-ply laminate subjected to longitudinal tension. Source: Sharifpour, F., Montesano, J. and Talreja, R., 2020. Assessing the effects of ply constraints on local stress states in cross-ply laminates containing manufacturing induced defects. *Composites Part B: Engineering*, *199*, 108227.

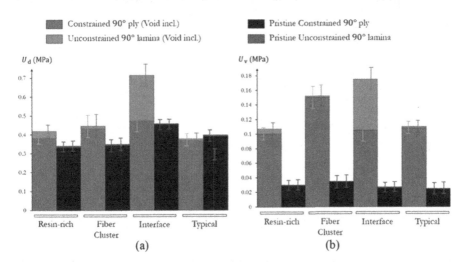

FIGURE 8.65 Distortional energy density (U_d) and dilatational energy density (U_v) compared for unconstrained and constrained plies. Source: Sharifpour, F., Montesano, J. and Talreja, R., 2020. Assessing the effects of ply constraints on local stress states in cross-ply laminates containing manufacturing induced defects. *Composites Part B: Engineering*, *199*, 108227.

Another study [106] followed the study [74] to further investigate the effects of manufacturing defects on the early stages of transverse crack formation under the constraint of adjacent undamaged plies in a cross-ply laminate. In this study, the presence of resin-rich pockets was included in addition to the nonuniform fiber distribution and voids. The micromechanical simulation was conducted on three-dimensional RVEs, and a different statistical procedure was implemented to generate fiber distributions than used previously [55, 56] as the degree of fiber distribution nonuniformity was not a subject of the study in [106]. While in previous works [55, 56] the post-yield response of the epoxy matrix was not viewed as of importance to the initiation of brittle cavitation as the first failure event under transverse tension, a pressure-sensitive yield response was included when axial shear was considered [80]. In [106] an inelastic model incorporating the effect of volumetric deformation in the inelastic regime was used to see if that would alter the failure analysis.

The previous conclusions concerning the initiation of brittle cavitation in regions of closely spaced fibers were confirmed in [106], while the implicit effect of including the inelastic behavior of epoxy was found to alter the locations and number of potential brittle cavitation sites. Additionally, as also indicated in previous two-dimensional simulations [55, 56], the importance of including thermal cooldown in the local stress analysis regarding dilatational energy density reaching criticality was verified.

Interesting insight into the effects of the ply constraint on the dilatational and distortional energy densities was gained in [106]. Figure 8.66 shows the increase of these quantities with the applied mechanical strain at points that reached their

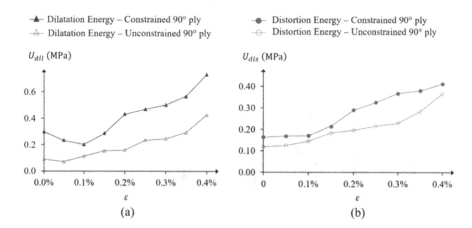

FIGURE 8.66 (a) Variation of dilatational energy density for a cavitated element in the constrained and unconstrained 90° ply and (b) variation of distortional energy density for an inelastically damaged element in the constrained and unconstrained 90° ply. Source: Sharifpour, F., Montesano, J. and Talreja, R., 2022. Micromechanical assessment of local failure mechanisms and early-stage ply crack formation in cross-ply laminates. *Composites Science and Technology*, 220, 109286.

criticality. In Figure 8.66(a) a significant effect of the ply constraint is seen on the dilatational energy density while the effect on the distortional energy density is not as high. Also noteworthy is the higher dilatational energy density at thermal cooldown (0% mechanical strain) in the constrained state than in the unconstrained state. This would imply earlier initiation of brittle cavitation in a cross-ply laminate than in a UD composite.

The simulation of a resin pocket is illustrated in Figure 8.67 [106]. As shown, two resin pockets of 10- and 25-μm size were considered and compared with the baseline case of no resin pockets. The effect of having a resin pocket is shown in

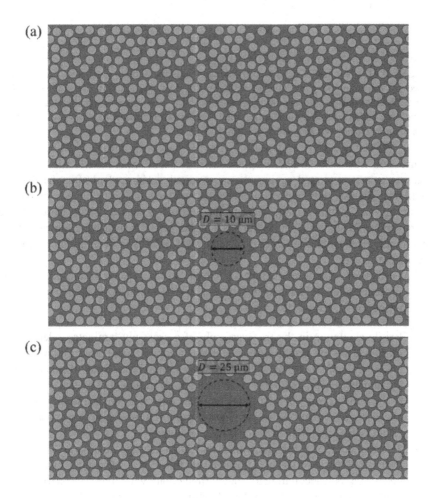

FIGURE 8.67   Representation of the nonuniform fiber distribution in the 90° ply within the laminate with: (a) No resin pocket (i.e., baseline case), (b) 10 μm resin pocket, (c) 25 μm resin pocket. Source: Sharifpour, F., Montesano, J. and Talreja, R., 2022. Micromechanical assessment of local failure mechanisms and early-stage ply crack formation in cross-ply laminates. *Composites Science and Technology, 220,* 109286.

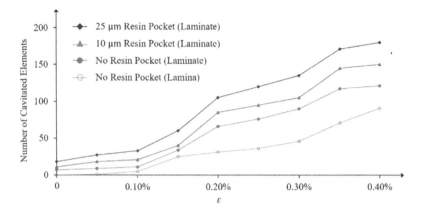

FIGURE 8.68 Number of cavitation incidents versus the applied mechanical strain influenced by the manufacturing-induced resin pocket presence and ply constraining effect. Source: Sharifpour, F., Montesano, J. and Talreja, R., 2022. Micromechanical assessment of local failure mechanisms and early-stage ply crack formation in cross-ply laminates. *Composites Science and Technology, 220*, 109286

Figure 8.68 where the number of positions in the RVE where the brittle cavitation criterion is met is plotted against the mechanical strain applied after the thermal cooldown. As seen, the number of cavitation incidents increases already by the ply constraint, and further enhancement in the number is caused by the presence of a resin pocket. The larger resin pocket has a higher effect in furthering the cavitation, presumably because of the larger displacement it induces in bringing the neighboring fibers together.

The effects of constraining effects on a transverse ply within a cross-ply laminate in the presence of manufacturing defects can now be summarized based on the studies reported in [74, 106] as follows.

- Brittle cavitation in the transverse ply of a cross-ply laminate under longitudinal tension occurs earlier (at a lower applied axial strain) than in an unconstrained UD composite. It is the first failure event of significance as it presumably results in fiber–matrix debonding. Initiation of matrix yielding in the transverse ply of a cross-ply laminate occurs as the next failure event, and it is also enhanced by the ply constraint. However, the ply constraint effect is less than in the case of brittle cavitation.
- Thermal cooldown enhances stress triaxiality close to fiber surfaces in the transverse ply of a cross-ply laminate, and this effect is also aided by the ply constraint. Not accounting for the thermal cooldown will give an erroneous stress state when calculating the dilatational energy density.
- While nonuniform fiber distribution is an important manufacturing defect in an analysis of the early stages of failure, matrix voids and resin-rich zones are also important in initiating first failure events, both in unconstrained and constrained states of UD composites.

8.5.2 PROGRESSION OF INTRALAMINAR FAILURE EVENTS IN LAMINATES

The transverse crack formation process in unconstrained UD composites was described above in Section 8.4.3.4 where the effects of the presence of voids in the matrix were also discussed. Once a transverse crack forms in a constrained ply of a cross-ply laminate, it grows unstably until it is arrested at the interfaces between the cracking ply and the adjacent ply. As described in Chapter 6, Section 6.2.1, more transverse cracks then form non-interactively, i.e., independently of the other transverse cracks, at first and at some point, under increasing axial tension on the laminate the transverse cracks interact, shielding mutual stress fields, and thereby reducing the rate of progression (multiple crack formation) process. This process has been studied extensively and its modeling has been performed by different approaches without considering manufacturing defects, see, e.g., [107, 108]. Figure 8.69 illustrates the local in-plane stress field resulting from a globally applied axial tensile force P on the laminate. After the first stage of transverse crack formation is completed, the in-plane stresses regarding the cracking ply as a homogeneous orthotropic layer govern the growth of the transverse crack. Since the transverse cracks grow unstably, the analysis of failure is focused on the crack multiplication process (i.e., formation of new fully grown cracks), as illustrated in Figure 8.70.

The work performed to create new N cracks is calculated as the work to close N cracks in an array of $2N$ cracks [107]. This requires knowing the (normal and shear) stresses on the crack planes and the crack surface displacements. Leaving the details to [107], it was found that the work on crack sliding displacements could be neglected in comparison to work on crack opening displacement (COD). Figure 8.71 shows the COD averaged over the crack length and normalized with a reference displacement for 90° cracks in $[0/90]_s$ and $[0/90/\mp45]_s$ laminates plotted against the crack density. The effect of crack shielding, i.e., stress perturbation from neighboring cracks, is clear, as seen in the COD reduction with increasing crack density (or reducing crack spacing).

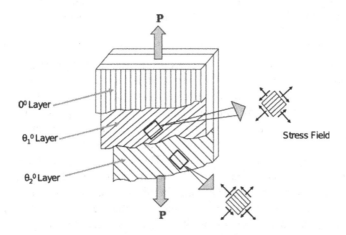

FIGURE 8.69 In-plane stress field in off-axis plies of a laminate under axial tensile load.

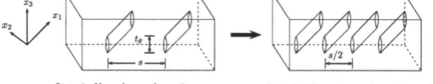

State 1: N cracks, crack spacing $= s$ State 2: $2N$ cracks, crack spacing $= s/2$

FIGURE 8.70 Failure even progression as multiplication of transverse crack formation process. Source: Singh, C.V. and Talreja, R., 2010. Evolution of ply cracks in multidirectional composite laminates. *International Journal of Solids and Structures*, 47(10), 1338–1349.

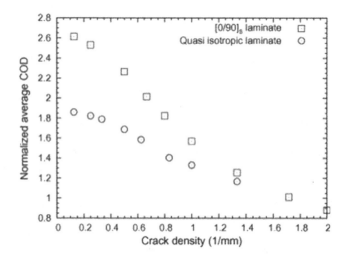

FIGURE 8.71 Variation of normalized average COD for 90°-cracks with crack density for [0/90]$_s$ and [0/90/∓45]$_s$ laminates. Source: Singh, C.V. and Talreja, R., 2010. Evolution of ply cracks in multidirectional composite laminates. *International Journal of Solids and Structures*, 47(10), 1338–1349.

The transverse crack evolution process for the [0/90]$_s$ and [0/90/∓45]$_s$ laminates is clarified by the experimental data reported in [109] and shown in Figure 8.72. As seen there, the crack multiplication process starts earlier for [0/90]$_s$ laminate compared to the [0/90/∓45]$_s$ laminate because of the lower ply constraint, as also indicated by the higher calculated COD for the [0/90]$_s$ laminate (Figure 8.71). As the external force is applied, the crack density in both laminates increases, approaching a saturation state of constant crack density. The initiation of crack formation in 45° plies, as observed experimentally, is indicated in the figure, verifying that the crack surface displacements are relatively small in this orientation, and therefore the work needed to create these cracks requires a higher applied load.

The numerical stress analysis in [107] was done for interacting cracks and is therefore valid when the crack density has reached a certain value where the crack

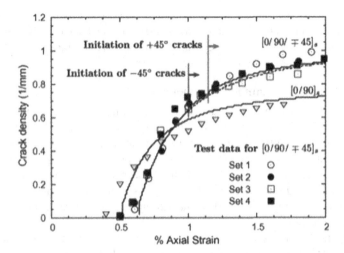

FIGURE 8.72 Evolution of 90°-crack density in [0/90]ₛ and [0/90/∓45]ₛ laminates. Source: Tong, J., Guild, F.J., Ogin, S.L. and Smith, P.A., 1997. On matrix crack growth in quasi-isotropic laminates—I. Experimental investigation. *Composites Science and Technology*, 57(11), 1527–1535

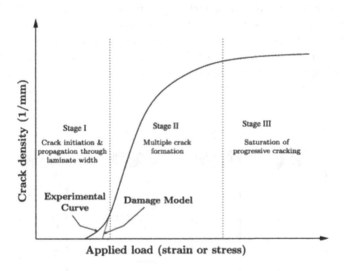

FIGURE 8.73 Stages in the evolution of ply cracking in laminates. Source: Singh, C.V. and Talreja, R., 2010. Evolution of ply cracks in multidirectional composite laminates. *International Journal of Solids and Structures*, 47(10), 1338–1349.

spacing is uniform. Experimental data shows, however, that in the early stage of the crack multiplication process, when the cracks are forming independently of other cracks, the average crack density has a different behavior than what is predicted by [107]. This is illustrated in Figure 8.73 where the stages of the crack density evolution process are indicated. As indicated, Stage I is obtained experimentally but not predicted by a crack interaction analysis. The next stage of interactive crack multiplication is for convenience separated into Stage II and Stage III, the latter indicating the crack density approaching a saturation state.

8.5.2.1 Effects of Manufacturing Defects

Figure 8.74 illustrates how the transverse ply cracking progresses with the applied axial tension on a cross-ply laminate. This is a typical scenario and indicates that the early stage of the multiple cracking process is random (non-interactive) where the cracks form at random locations in the transverse ply. This corresponds to Stage I in Figure 8.73. As the loading is increased, the crack spacing becomes increasingly uniform (Stage II), and the rate at which new cracks form between pre-existing cracks reduces. The final stage (Stage III) is the saturation state where the crack spacing is essentially constant.

The models such as that presented in [107] treat interactive crack multiplication (Figure 8.70) and successfully predict the crack evolution process in Stages II and III. This suggests that the effects of manufacturing defects are not significant in these stages. To understand how manufacturing defects affect the entire crack evolution process, an experimental study was conducted to systematically investigate the effects of defects by preparing cross-ply laminates in an autoclave with different deviations from the specified manufacturing process [110]. Specimens from the panels so produced were subjected to axial tension. The experimental results on observations concerning the crack formation induced by voids were reviewed in Chapter 6, Section 6.2.1. The transverse crack densities in the three cases of defective specimens were compared with the case of normal specimens, i.e., those manufactured without deviating from the recommended curing process. As expected, the crack

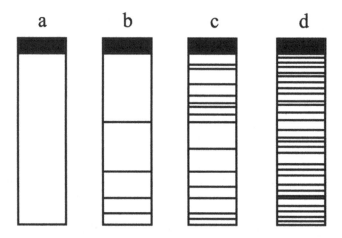

FIGURE 8.74 Progression of multiple transverse cracking (horizontal lines) with increasing axial tension on a cross-ply laminate. (a) No load, (b) Stage I, (c) Stage II, and (d) Stage III (refer to Figure 8.73). Source: Talreja R. and Singh, C.V., 2012. Cambridge University Press.

densities were found to be higher in the defective cases than in the normal case. Here we shall describe the modeling approach used in [110] to account for the effects of manufacturing defects on the crack evolution process.

Motivated by the observed random locations of transverse cracks in Stage I, the transverse strength of the 90° plies was assumed to be nonuniform due to the presence of defects. The random distribution of defects in these plies can thus indirectly be accounted for by assuming the transverse strength σ_s to be given by Weibull distribution,

$$P_s\left(\sigma_s\right) = 1 - \exp\left[-\left(\sigma_s / \sigma_0\right)^m\right] \qquad (8.16)$$

where σ_0 and m are the scale and shape parameters, respectively, of the distribution. It is noted that the tensile strength in Equation (8.16) is not that of a stand-alone (unconstrained) ply but of a constrained ply within the cross-ply laminate.

Figure 8.75 illustrates the varying crack spacing L over which the axial stress is calculated for a carbon–epoxy laminate. When this spacing is large, the perturbation in the axial stress caused by one crack does not interact with perturbations by the other cracks, as the distance over which the perturbation lasts is much less than the crack spacing (St. Venant's principle). The stress variation, normalized by its pre-crack value, is shown in Figure 8.76 for two large and small crack spacings, L = 20 mm and L = 2 mm, as examples of Stage I and Stage II cracking processes, respectively.

Random numbers are drawn to assign strength values to the discrete elements of the 90° ply according to the probability distribution, Equation 8.16. As seen in Figure 8.76(a), the local axial stress for the non-interactive crack spacing is nearly the same as the ply stress σ_{xx} in the uncracked 90° plies, except in a short distance close to the crack planes. This stress can be calculated with good accuracy by using the classical laminate theory. The crack density, i.e., the number of cracks per unit length, which is the same as the number of failed elements per unit length, is calculated from the failure probability of elements according to Equation (8.16) as,

$$\rho = \frac{1}{l_0} P_s\left(\sigma_s \leq \sigma_{xx}\right) \qquad (8.17)$$

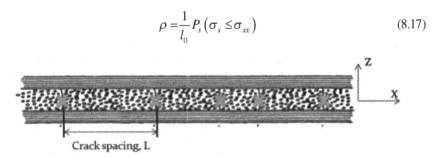

Crack spacing, L

FIGURE 8.75 The 90° plies divided by the cracks into blocks with varied lengths. Source: Huang, Y., Varna, J. and Talreja, R., 2014. Statistical methodology for assessing manufacturing quality related to transverse cracking in cross ply laminates. *Composites Science and Technology*, 95, 100–106.

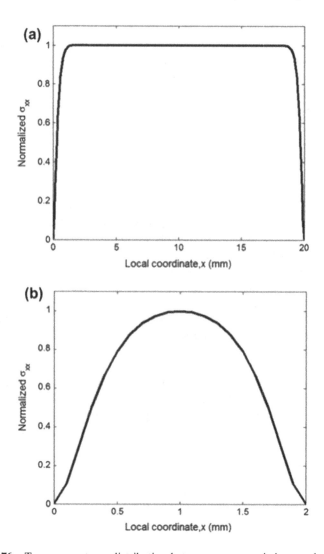

FIGURE 8.76 Transverse stress distribution between two preexisting cracks. The stress is normalized by the maximum value; (a) $L = 20$ mm and (b) $L = 2$ mm. Source: Huang, Y., Varna, J. and Talreja, R., 2014. Statistical methodology for assessing manufacturing quality related to transverse cracking in cross ply laminates. *Composites Science and Technology*, *95*, 100–106.

From the measured crack densities in specimens manufactured by a standard procedure (V_P), i.e., with vacuum and pressure applied, and three non-standard procedures (NV_NP), (NV_P), and (V_NP), where N stands for non-application, the Weibull parameters in Equation (8.16) were estimated. Using these parameters in the failure simulation procedure, the probabilities of failure for all four cases were calculated and are shown in Figure 8.77. The upper figure, Figure 8.77(a) is for the

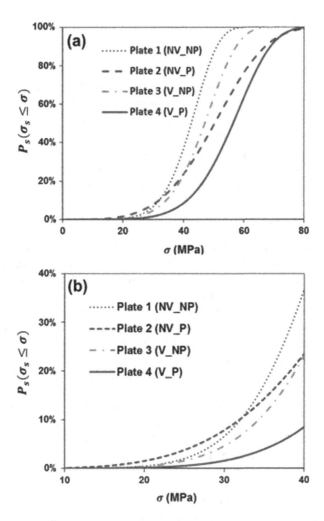

FIGURE 8.77 The statistical distribution of static transverse strength of four plates manufactured by different processes: (a) full range and (b) lower range. Source: Huang, Y., Varna, J. and Talreja, R., 2014. Statistical methodology for assessing manufacturing quality related to transverse cracking in cross ply laminates. *Composites Science and Technology*, 95, 100–106.

full range of the transverse cracking process, which includes the interactive part of the cracking process (see [110] for details). The lower range, shown in Figure 8.77 (b) is for the non-interactive part and represents the effects of manufacturing defects. As seen here, the standard (non-defective) case has the lowest probability of crack formation, while the failure probability is higher for the three cases of non-standard manufacturing.

It is interesting to note that plate 2 (NV_P), not plate 1 (NV_NP), has the highest fraction of the probability distribution in the lower strength range, i.e., $P_s(\sigma_s \leqslant 30$

MPa). A probable reason is that in plate 2 (NV_P) with only pressure applied during the curing process, gas bubbles gather driven by the applied pressure. Lacking the vacuum system, the gathered gas forms large voids, which significantly reduce the local strength. In plate 1 (NV_NP), although the gas cannot be removed without applying a vacuum, it would not accumulate and form large voids, as it would in plate 2 (NV_P), because no pressure is applied. Instead, the gas forms small voids, which reduce the strength of the plate moderately and more broadly. This argument is supported by the higher fraction of relatively low strength (~35 MPa) for plate 1 (NV_NP). By comparing plate 3 (V_NP) to plate 1 (NV_NP) and plate 2 (NV_P), it can also be seen that applying a vacuum during the curing process is more effective than applying pressure to reduce defects that are critical for transverse cracking.

8.5.3 DELAMINATION

When transverse cracks formed in plies of a laminate subjected to in-plane loading reach interfaces with adjacent plies, the intense field carried by the cracks inevitably fails (debonds) the interfaces locally. This local delamination is usually small and can go undetected except at the free edges of rectangular specimens if the edges are properly polished. In fact, the delamination observed at free edges can be caused by the three-dimensional stress state existing locally (extending a short distance into the laminate width) caused by the traction-free free edges. The free-edge delamination is by itself of interest for analyzing failure from holes, cutouts, ply drops, etc. Our interest here is to analyze the effects of manufacturing-induced delaminations that are usually distributed in the volume of a laminate and can cause failure.

Delaminations caused by manufacturing processes were reviewed in Chapter 6, Section 6.4. As described there, such delamination defects can be classified as: a) Delamination at free edges induced by machining (cutting, drilling, etc.), b) delamination at curved geometries induced by inadequate resin wetting and curing, and c) delamination due to incorrect molding.

8.5.3.1 Delamination Fracture

The presence of delamination can cause failure under service loading depending on the size and shape of the delamination crack and the loading mode. If a delamination is at an interface between two differently oriented plies (dissimilar materials), then under in-plane loading the delamination crack will grow in a mixed-mode. Several methods for calculating the energy release rates (ERRs) for interface cracks have been proposed. In [111] an exact formulation of ERR based on the J-integral was derived and a decomposition in mode I and mode II ERR was proposed. The virtual crack closure technique (VCCT) was used in [112, 113] to evaluate the total and individual ERRs along the delamination front. Beyond these early works, significant developments have taken place that are reviewed in [114]. The review also covers cohesive zone models and compares these to VCCT-based approaches. If the manufacturing-induced delamination contains bridging of the two crack planes, possibly due to fibers crossing those planes, then using cohesive zone models can be justified. The crack bridging has been treated by cohesive zone models in [115, 116].

Assessing failure from pre-existing manufacturing-induced delamination requires the use of a fracture criterion for mixed-mode crack growth. At least four such criteria have been proposed, all described as "laws" [114]. Each criterion is based on assumptions inspired by certain experimental data and/or convenience/flexibility of curve-fitting data. As is common in many deformation and failure problems (plasticity, fatigue, fracture), resorting to power laws provides a convenient way to develop engineering approaches. In any case, finding the fracture toughness, i.e., the critical ERR, is a problem coupled with calculating the total ERR and its mode components. For this purpose, several standard specimens have been used. For mode I fracture toughness the most common specimen is the double cantilever beam (DCB) specimen using a UD composite. For pure mode II fracture toughness, the same UD composite can be used as an end-notch flexure specimen, while a mixed-mode bending specimen is suitable for the determination of crack growth under the combined effect of the two modes [117, 118]. A schematic representation of the three specimen types is depicted in Figure 8.78 [118].

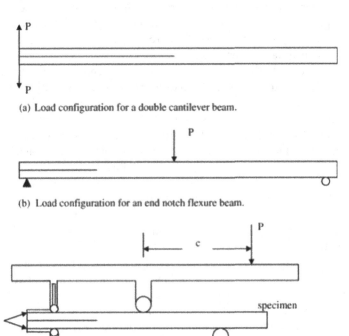

(a) Load configuration for a double cantilever beam.

(b) Load configuration for an end notch flexure beam.

(c) Mixed-mode bending specimen mounted on the loading apparatus.

FIGURE 8.78   Schematic representation of mode I, mode II, and mixed-mode fracture test configurations. Source: Mathews, M.J. and Swanson, S.R., 2007. Characterization of the interlaminar fracture toughness of a laminated carbon/epoxy composite. *Composites science and technology, 67*(7–8), 1489–1498.

In fracture mechanics, it is assumed that resistance to fracture is given by the material in the process zone ahead of the crack front. This assumption is verified in cases of crack growth in pure modes. However, in mixed-mode crack growth, the self-similarity of crack growth does not always hold, as was observed in the study [119] where the mixed-mode bending specimens were tested with cracks placed between two plies of different orientations. The apparent variation of the critical value of the total ERR with the angle of mismatch between the ply orientations found in [119] is perhaps an indication of the inapplicability of LEFM-based analysis for mixed-mode failure rather than a true reflection of the fracture resistance. The study [120] thoroughly examined the fracture toughness in multidirectional laminates and found that the mode I toughness varied slightly with interface layup and direction of delamination growth. The underlying cause of this variation appeared to be the deviation of the crack growth path from the plane of the pre-crack. This lack of self-similar crack growth was found to be more severe for mode II cracking resulting in large variation of the fracture toughness in what should be a material property. The large number of proposed empirical relationships for mixed-mode fracture toughness in the literature raises doubt in viewing fracture toughness for this case as a material property. Only for mode I fracture toughness, it seems reasonable to assume this to be a material property.

It is conceivable that the toughness of the polymer matrix contributes to the delamination fracture toughness. To examine if there is a relationship between the two, one study [121] varied the ductility (tensile elongation to failure) of epoxy and tested carbon/epoxy composites for mode I and mode II delamination toughness. It was found that the increased ductility in epoxy increased its mode I fracture toughness and increased mode I and mode II delamination toughness values as well. While these increases were at significant rates up to the epoxy fracture toughness of 700 J/m^2, beyond this value the beneficial effect on delamination toughness dropped off sharply. It must be recognized that the stress state of the matrix in the thin plane of delamination is not the same as in its bulk form. Thus, the process zone ahead of the crack front that provides fracture resistance in the two cases is also not the same.

Defective manufacturing can also result in unwetted fibers leading to imperfect fiber–matrix bonding. To investigate how poor bonding could affect delamination toughness, one study [122] looked at the effect of the bonding efficiency by applying four different wet oxidative surface treatments to carbon fibers before embedding them in a thermosetting resin and a thermoplastic system. Delamination fracture toughness values were determined for mode I, mode II, and mixed-mode cases. It was found that all values increased for both matrix materials with increasing surface treatment. This suggests that fiber–matrix debonding ahead of the delamination front aids in crack propagation, as would be expected.

8.5.3.2 Effect of Voids

The effect of voids on delamination fracture toughness has been studied experimentally in [123, 124, 125]. While voids were induced in the composites by manipulating the curing process, the distribution of voids achieved was highly nonuniform. The influence on delamination toughness by the presence of voids in the interface could not be determined with confidence.

As noted above, the resistance to delamination growth is lowest in mode I. A systematic analytical and computational study was performed in [126] using a DCB specimen. It is expected that, if voids are trapped within the interface, then the resistance to crack growth can be affected. Equivalently, one can view the ERR of the crack tip in the DCB specimen to be enhanced by the voids. Figure 8.79 shows a schematic representation of the DCB specimen with a void whose center lies at a distance d ahead of the crack of length a. For a void of circular geometry, the radius R is specified, while for an elliptical void, an aspect ratio is additionally given.

Figure 8.80 shows the computed effect of the presence of a void on G_I normalized by its non-void value for a circular and elliptical void, as a function of the distance of the void from the crack tip. The analytical predictions are compared

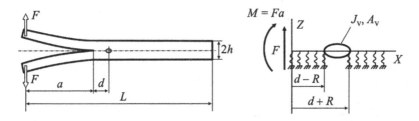

FIGURE 8.79 A schematic of a DCB specimen with a void ahead of the crack tip (left) and the beam-on-elastic-foundation model (right). A_v and J_v are the cross-sectional area and the moment of inertia of the void, respectively. Source: Ricotta, M., Quaresimin, M. and Talreja, R., 2008. Mode I strain energy release rate in composite laminates in the presence of voids. *Composites Science and Technology*, 68(13), 2616–2623.

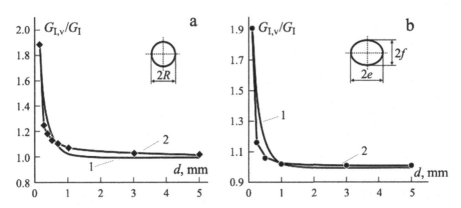

FIGURE 8.80 The effect of circular ($R = 0.1$ mm) (a) and elliptical ($e = 0.1$ mm, $f = 0.05$ mm) and (b) voids on the mode I SERR with their distance d from the crack tip according to the analytical model (1) and FE results (2). Source: Ricotta, M., Quaresimin, M. and Talreja, R., 2008. Mode I strain energy release rate in composite laminates in the presence of voids. *Composites Science and Technology*, 68(13), 2616–2623.

with the finite element (FE) computations in the figure and show good agreement. Having validated the analytical model, results for different cases were computed. Figure 8.81 shows the effect of void size and shape illustrated by considering elliptical voids of different aspect ratios. Both the void size and the shape effects on the ERR are significant. Figure 8.82 illustrates the effect of having multiple voids ahead of the crack tip. As seen, the effect of the nearest void has the greatest influence on the ERR, while adding more voids has an increasingly less effect. Finally, when the position of the first void is fixed, as shown in Figure 8.82 (right), the effect of voids beyond the nearest one from the crack tip shows a maximum at a short distance from the first void.

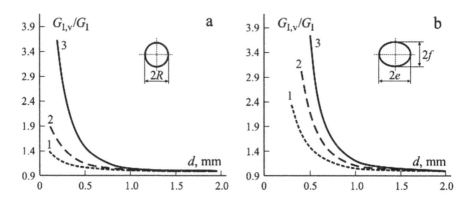

FIGURE 8.81   Effect of void size (a) and shape (b) on the mode I ERR. (a) $R = 0.05$ (1), 0.1 (2) and (b) $f = 0.1$ mm; $e/f = 2$ (1), 3 (2), and 4 (3) 0.2 mm (3). Source: Ricotta, M., Quaresimin, M. and Talreja, R., 2008. Mode I strain energy release rate in composite laminates in the presence of voids. *Composites Science and Technology*, 68(13), 2616–2623.

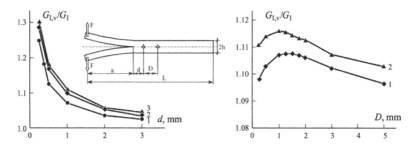

FIGURE 8.82   Left: Effect of multiple aligned voids, with $R = 0.1$ mm and mutual distance between them $D = 2$ mm, on the mode I SERR in the cases of one (1), two (2), and three (3) voids. Right: Effect of two (1) and three (2) aligned voids with $R = 0.1$ mm on the mode I ERR as the distance D of additional voids from the first void. The distance of the first void from the crack tip is $d = 1$ mm. [126]. Source: Ricotta, M., Quaresimin, M. and Talreja, R., 2008. Mode I strain energy release rate in composite laminates in the presence of voids. *Composites Science and Technology*, 68(13), 2616–2623.

The main results of the parametric study of the effects of voids on delamination fracture toughness can be summarized as follows.

- In the presence of voids, the mode I ERR value increases as the dimension of the void increases and with the decrease in the distance between the void and crack tip.
- The void shape plays an important role: circular and elliptical voids are equally critical when the radius of the circular void and the half-major axis of the elliptical void are the same. On the other hand, elliptical voids are much more critical when the minor half-axis of the voids equals the radius of the circular void. Moreover, ERR values for elliptical voids increase significantly with the aspect ratio.
- With an increasing number of aligned voids, the influence on the mode I ERR is controlled by the first two voids.
- The effect of voids beyond the nearest one from the crack tip shows a maximum at a short distance from the first void.

Finally, in experimental work, it is difficult to distribute voids in a controlled manner so that their effects on the delamination fracture can be studied. For engineering design purposes, therefore, the parametric study described here can allow the assessment of the effects of voids more effectively.

8.5.3.3 Delamination Buckling

Failure of a laminate with pre-existing manufacturing-induced delamination can be caused by a compressive in-plane load. Figure 8.83 [127] illustrates the possible local and global buckling loads. The modeling work to predict failure by unstable growth of delamination goes back to 1981 [128] where a delaminating beam-column was considered, and the growth of delamination was analyzed by the fracture mechanics-based ERR criterion. Later, a more accurate beam-plate model was used [129] in which the location of delamination was considered as a parameter. A finite element approach [130] was presented with more geometrical parameters. Multiple delaminations have also been considered [131, 132].

An extensive theoretical and computational study [133, 134] provided the following useful conclusions.

- The initial buckling load of delaminated composite laminates is often much lower than the global buckling load. To determine the load capacity of the delaminated composite laminates, post-buckling analysis is necessary.
- The delamination propagation has a significant effect on the post-buckling response and the strength of the laminates.
- Unstable delamination propagation is often caused by mode I fracture while the mode II fracture often leads to stable delamination growth as far as all the applications considered here are concerned.
- Due to the interaction between the delaminations, the post-buckling response of the composite laminate with multiple delaminations can be quite different from that with a single delamination.

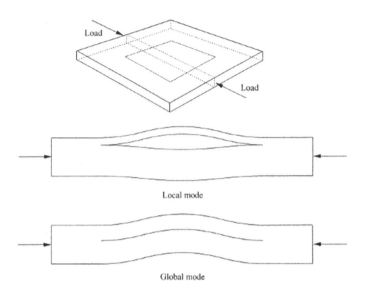

FIGURE 8.83 Local and global buckling modes of a panel with embedded delamination. Source: Short, G.J., Guild, F.J. and Pavier, M.J., 2001. The effect of delamination geometry on the compressive failure of composite laminates. *Composites Science and Technology*, *61*(14), 2075–2086.

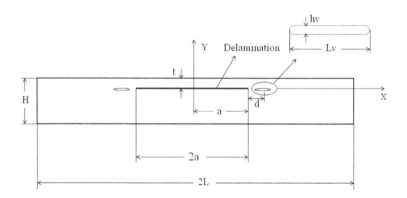

FIGURE 8.84 The model to analyze post-buckling behavior of a delaminated plate with voids ahead of the delamination crack front. Source: Zhuang, L. and Talreja, R., 2014. Effects of voids on postbuckling delamination growth in unidirectional composites. *International Journal of Solids and Structures*, *51*(5), 936–944.

Guided by these findings, a study was conducted to examine the postbuckling behavior of a delaminated plate in the presence of voids [135]. The model considered is depicted in Figure 8.84.

A parametric study was conducted with the size of voids and their distance from the delamination front was taken as the parameters. A nonlinear FE analysis was performed in which the ERR of the delamination crack was computed by the VCCT. The computations of the total ERR were validated against an analytical prediction by [128] and a J-integral calculation. The VCCT results compared perfectly well with the analytical model and had negligible differences with respect to the J-integral method. The mode I and mode II ERRs were then computed with and without voids and are shown in Figure 8.85 for one case [135]. As seen in the figure, in the opening phase of the delamination front, i.e., before G_I reaches its maximum value, the presence of void decreases G_I slightly. The effect of void on G_I becomes more significant during the closing tendency of the delamination, when the presence of void enhances G_I. However, this effect is of no practical interest since failure from delamination will occur at or before reaching the maximum G_I value. Unlike G_I, the presence of a void increases G_{II} regardless of whether the delamination front is opening or closing. Meanwhile, note that for the same *applied load*, the G_{II} value is much larger than G_I, which is due to the proximity of the delamination to the plate surface. As a result, the total ERR G_T is dominated by G_{II}, and the effect of void on G_T is expected to follow the same trend as that for G_{II}.

The parametric study [135] summarized the following findings.

- The presence of void shows negligible effect on the buckling displacement of the part containing delamination.
- The effect of void on G_I shows a complex trend. In the opening phase of the delamination front, until G_I attains its maximum value, different sizes and locations of the void have different effects on G_I and depending on the composite stiffness ratio and thickness of the delaminated part of the plate, the presence of void may decrease or increase the maximum G_I.
- In the presence of voids, G_{II} increases with increasing void size and with decreasing distance from the delamination front.
- With all other parameters the same, the presence of voids is likely to be more detrimental for a structure undergoing local buckling with a thicker delaminated plate.

As an overall conclusion, the results illustrate that the size, shape, and position of the voids have a specific influence that cannot be properly captured in the void volume fraction. Thus, placing a threshold on the void volume fraction as an acceptance/rejection criterion in design is inadequate.

8.5.4 FINAL FAILURE

Laminates are designed to have plies oriented such that sufficient plies are aligned along the major load direction. Examples are spars in wind turbine blades and in aircraft wings. The final failure in that case is caused by the breakage of fibers under the major load. Modeling of failure under longitudinal tension was considered above in Section 8.4.1 and under longitudinal compression in Section 8.4.2 for UD composites. When placed within a laminate, the final failure modeling of the plies must additionally consider the changes in the local stress states brought about by

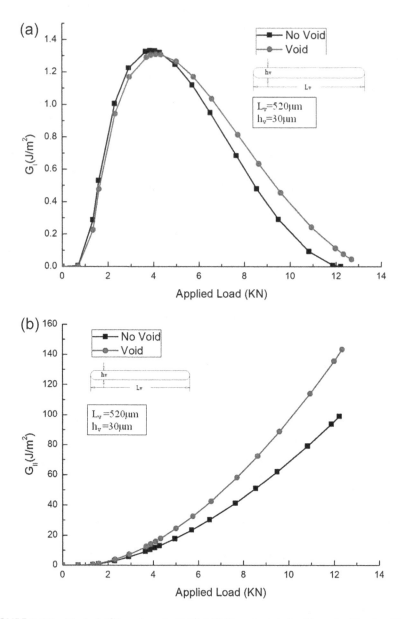

FIGURE 8.85 Mode I (G_I) and mode II (G_{II}) ERRs calculated with and without voids for the void dimensions shown placed at distance d = 0.4 mm from the delamination crack tip. Source: Zhuang, L. and Talreja, R., 2014. Effects of voids on postbuckling delamination growth in unidirectional composites. *International Journal of Solids and Structures*, *51*(5), 936–944.

the presence of the adjacent (constraining) plies. In other words, combined loading must be considered with the primary in-plane loading being longitudinal tension or compression. Furthermore, if the adjacent plies undergo intralaminar cracking, then the local stress enhancement induced by the crack fronts at the ply interfaces must also be considered. This consideration should include the local delamination at the crack fronts if it is significant.

8.5.4.1 Longitudinal Tension Combined with In-Plane Shear

Under longitudinal tension of a laminate such as $[0/\pm\theta]_s$ the θ-plies will induce in-plane shear in the entire $0°$ plies unless the θ-plies undergo intralaminar cracking, in which case, the effect will be local at the crack fronts. Because of the combined longitudinal tension and in-plane shear, the fiber breakage and the fiber–matrix debonding emanating from the broken fiber ends will be affected. The process of fiber breakage and fiber-matrix debonding was treated in Section 8.4.1 above for UD composites under longitudinal tension. This process must now be considered under the altered fiber and matrix stresses. The statistical fiber failure process is not expected to change under the altered stresses in the longitudinal plies as the fiber failure will still be governed by the axial tensile stress in the fibers. The fiber–matrix debonding that follows from the broken fiber ends will, however, no longer be axisymmetric. Also, the kink out of the debond (Figure 8.9) toward the neighboring fiber will be affected by the presence of in-plane shear.

In Section 8.4.1 above, the case of a pre-existing broken fiber was analyzed (Scenario 1) under longitudinal tension. It was found that a matrix crack is emanated from the broken fiber end. When an in-plane shear is superimposed, the direction in which the matrix crack grows will no longer be normal to the fiber axis but will tilt to become normal to the maximum local tensile stress. This will have consequences on the debonding of the neighboring fiber and therefore on its probability of failure.

8.5.4.2 Effect of Intralaminar Cracking in the Constraining Plies

If in a laminate such as $[0/\pm\theta]_s$ the θ-plies suffer intralaminar cracking, then the intense stress field carried by the cracks at their fronts will induce high local stress enhancement on the fibers in the $0°$ plies. The stresses at the crack fronts will also induce local delamination, the extent and shape of which will depend on the angle θ. To focus on these local effects, studies have often considered $[0/90]_s$ cross-ply laminates. Among the earliest studies to investigate the fiber breakage in $0°$ plies with and without the $90°$ plies was in [136]. This study separated $0°$ plies from the $90°$ plies intermittently during axial tension and from SEM images counted the number of fiber breaks over a selected planar area. The number density of the observed fiber breaks is compared for the two cases – with and without the constraining $90°$ plies with transverse cracks in Figure 8.86. Until approximately 0.7% strain the fiber breakage process is about the same in both cases, but beyond this the number of fiber breaks increases in the cross-ply laminate. The observations reported in [136] suggest that the increased number is due to the fiber breaks close to the region of the transverse crack fronts intersecting the longitudinal plies. It is interesting to note that despite this the failure strain in both cases is still nearly the same.

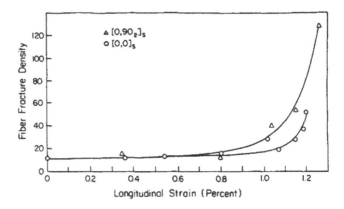

FIGURE 8.86   Number of fiber breaks per unit area plotted against the longitudinal strain in UD composite [0/0]$_s$ and in [0/90$_2$]$_s$ laminate. Source: Jamison, R.D., 1985. The role of micro-damage in tensile failure of graphite/epoxy laminates. *Composites Science and Technology, 24*, 83–99.

Observations reported in [136] also indicate that most of the increased number of fiber breaks in the longitudinal plies of the cross-ply laminate are isolated fiber breaks, also described as singlets. These singlets induce breakage of the neighboring intact fibers by a process often described as local load sharing [137] and are attributed to the stress concentration induced by a broken fiber on its intact neighbors. In [22] the process of local load sharing was analyzed in detail by accounting for fiber–matrix debonding from the broken fiber end of the kink out toward the intact fibers. The increase in the probability of failure of intact fibers over the affected length was calculated and found to be significant. Thus, the formation of doublets from singlets is a highly local phenomenon limited to the immediate proximity of intact fibers to a broken fiber end and would not be influenced significantly by the stress enhancement near the transverse crack fronts. The statistics concerning the occurrence of doublets and higher multiplets reported in [137] suggested that in longitudinal plies of cross-ply laminates, this occurrence was not more likely than in unconstrained UD composites. This may explain why the tensile failure strain in both cases was not significant.

8.5.4.3   Longitudinal Compression Combined with Shear

The modeling of failure of an unconstrained UD composite under longitudinal compression was described in Section 8.4.2 above where fiber misalignment and matrix voids were considered as manufacturing defects. When a UD composite is within a laminate, such as in [0/±θ]$_s$, a longitudinal compression load on the laminate will induce in-plane shear in addition to the axial compression in the 0° plies. This combined loading can be viewed as equivalent to applying an axial compression load on the UD composite at an off-axis angle. A study considered off-axis compression failure of a UD glass–epoxy composite [138] by the microbuckling model and kink-band model separately. In the microbuckling model [43], they regarded the small initial fiber misalignment angle and the small off-axis loading angle as two constant angles while the in-plane shear strain was taken to increase with the applied load. Thus, the in-plane shear stress–strain response was modeled as nonlinear. For the kink-band model [44], they approximated the critical load under off-axis compression by accounting only for the shear stress in the kink band and like in the microbuckling model added the small fiber misalignment angle and the small off-axis angle as the total fiber misalignment angle. The predictions by the two models were compared with test data for 5° and 10° off-axis loading and found to compare well with both model predictions. The authors recommended to use the kink-band model, described above in Section 8.4.2 (Equation 8.7) with the angle ϕ representing the fiber misalignment angle plus the off-axis angle. The fixed shear stress linear elastic limit τ_y and the corresponding strain γ appearing in Equation 8.7 were not used. Instead, the shear stress and shear strain were varied according to a nonlinear stress–strain relationship (curve-fit from experimental data) and the maximum critical compressive stress thus obtained was taken as the failure stress. This procedure can be relied upon for small values of off-axis angles, and therefore for relatively small in-plane shear stresses. For higher in-plane stresses, the fiber microbuckling model [43] and the kink-band model [44] may not apply as the in-plane shear stress itself may induce failure. Also, the transverse normal stress on the fibers in that case will also affect the failure process.

Later, another study [139] tested UD glass–epoxy specimens in compression at off-axis angles of up to 90° while recording the stress-strain response and observing the failure mechanism by SEM. The compressive strength is shown plotted in Figure 8.87. The mechanisms for fiber failure at low angles were observed to be

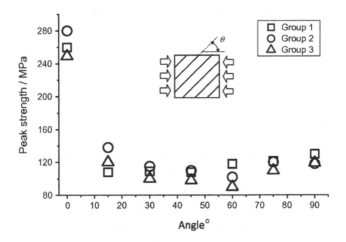

FIGURE 8.87 Variation of compressive strength with off-axis angle for a glass–epoxy UD composite. Source: Zhou, H.W., Yi, H.Y., Gui, L.L., Dai, G.M., Peng, R.D., Wang, H.W. and Mishnaevsky Jr, L., 2013. Compressive damage mechanism of GFRP composites under off-axis loading: experimental and numerical investigations. *Composites Part B: Engineering*, *55*, 119–127.

fiber microbuckling and kinking, while at large angles the failure occurs by shear cracking of the composite. These observations confirmed the limitations of using the fiber microbuckling model or the kink-band model to small off-axis angles, as noted above. For larger angles, up to 45°, the in-plane shear stress caused cracking of the plies along fibers. For even larger angles, the failure was essentially in the compression of the matrix.

An earlier study [140] tested $[\pm\theta]_s$ laminates in axial compression and found results similar to those reported for off-axis loading of UD composites [139], Figure 8.87, except at small angles, $\theta < 15°$, the interlaminar shear influenced the compression failure.

8.6 CONCLUSION

This chapter has focused on the physical modeling of failure in composite materials, which is at the core of the manufacturing sensitive design strategy described in Chapter 1. It began with a review of phenomenological failure theories to clarify why such theories are unable to properly account for manufacturing-induced defects. A mechanisms-based multiscale modeling strategy was then discussed. Following the observed failure mechanisms described in Chapter 5, this strategy was discussed systematically for failure in unidirectional composites under different individual loading modes. Quantification of defects as defect severity measures was described and incorporation of these measures in the failure models was discussed. The statistical descriptors of fiber distribution nonuniformity described in Chapter 4 were shown to be incorporated in the failure analysis. The effect of voids on fiber–matrix debonding was discussed. In this context, the use of cohesive zone modeling and phase field fracture concepts were examined. For laminates, the modeling of ply cracking evolution and delamination initiation and progression were discussed.

REFERENCES

1. Talreja, R., 2014. Assessment of the fundamentals of failure theories for composite materials. *Composites Science and Technology, 105*, pp. 190–201.
2. Talreja, R., 2016. Physical modelling of failure in composites. *Philosophical Transactions of the Royal Society A: Mathematical, Physical and Engineering Sciences, 374*(2071), 20150280.
3. Azzi, V.D. and Tsai, S.W., 1965. Anisotropic strength of composites: Investigation aimed at developing a theory applicable to laminated as well as unidirectional composites, employing simple material properties derived from unidirectional specimens alone. *Experimental Mechanics, 5*, pp. 283–288.
4. Hill, R., 1948. A theory of the yielding and plastic flow of anisotropic metals. *Proceedings of the Royal Society of London. Series A. Mathematical and Physical Sciences, 193*(1033), pp. 281–297.
5. Tsai, S.W. and Wu, E.M., 1971. A general theory of strength for anisotropic materials. *Journal of Composite Materials, 5*(1), pp. 58–80.
6. Gol'denblat, I.I. and Kopnov, V.A., 1965. Strength criteria for anisotropic materials. *Mecchanica IMZGA, Izvestia Academy, Nauk USSR, 6*, pp. 77–83.

7. Hashin, Z., 1980. Fatigue failure criteria for unidirectional fiber composites. *Journal of Applied Mechanics, 47*(4), pp. 329–334.
8. Puck, A. and Schürmann, H., 2002. Failure analysis of FRP laminates by means of physically based phenomenological models. *Composites Science and Technology, 62*(12–13), pp. 1633–1662.
9. Puck, A. and Schürmann, H., 2004. Failure analysis of FRP laminates by means of physically based phenomenological models. In M. Hinton, P.D. Soden, A.S. Kaddour (Eds.), *Failure Criteria in Fibre-Reinforced-Polymer Composites* (pp. 832–876). Amsterdam: Elsevier.
10. Christensen, R.M., 2019. Failure criteria for fiber composite materials, the astonishing sixty year search, definitive usable results. *Composites Science and Technology, 182*, 107718.
11. Gu, J., Chen, P., Su, L. and Li, K., 2021. A theoretical and experimental assessment of 3D macroscopic failure criteria for predicting pure inter-fiber fracture of transversely isotropic UD composites. *Composite Structures, 259*, 113466.
12. Davila, C.G. and Camanho, P., 2003. *Failure Criteria for FRP Laminates in Plane Stress*. NASA/TM-2003-212663. Hampton, VA: National Aeronautics and Space Administration
13. Hill, R., 1963. Elastic properties of reinforced solids: Some theoretical principles. *Journal of the Mechanics and Physics of Solids, 11*, pp. 357–372.
14. Kwon, Y.W., Allen, D.H. and Talreja, R., 2008. *Multiscale Modeling and Simulation of Composite Materials and Structures*, Vol. *47*. New York: Springer.
15. Aboudi, J., Arnold, S.M. and Bednarcyk, B.A., 2013. *Micromechanics of Composite Materials: A Generalized Multiscale Analysis Approach*. Amsterdam: Elsevier.
16. LLorca, J., González, C., Molina-Aldareguía, J.M., Segurado, J., Seltzer, R., Sket, F., Rodríguez, M., Sádaba, S., Muñoz, R. and Canal, L.P., 2011. Multiscale modeling of composite materials: A roadmap towards virtual testing. *Advanced Materials, 23*(44), pp. 5130–5147.
17. Bunsell, A., Gorbatikh, L., Morton, H., Pimenta, S., Sinclair, I., Spearing, M., Swolfs, Y. and Thionnet, A., 2018. Benchmarking of strength models for unidirectional composites under longitudinal tension. *Composites Part A: Applied Science and Manufacturing, 111*, pp. 138–150.
18. Hedgepeth, J.M., 1961. *Stress Concentrations in Filamentary Structures*. NASA Technical Note D-882.
19. Hedgepeth, J.M. and Van Dyke, P., 1967. Local stress concentrations in imperfect filamentary composite materials. *Journal of Composite Materials, 1*(3), pp. 294–309.
20. Okabe, T. and Takeda, N., 2002. Elastoplastic shear-lag analysis of single-fiber composites and strength prediction of unidirectional multi-fiber composites. *Composites Part A: Applied Science and Manufacturing, 33*(10), pp. 1327–1335.
21. Ohno, N., Okabe, S. and Okabe, T., 2004. Stress concentrations near a fiber break in unidirectional composites with interfacial slip and matrix yielding. *International Journal of Solids and Structures, 41*(16–17), pp. 4263–4277.
22. Zhuang, L., Talreja, R. and Varna, J., 2016. Tensile failure of unidirectional composites from a local fracture plane. *Composites Science and Technology, 133*, pp. 119–127.
23. Swolfs, Y., McMeeking, R.M., Verpoest, I. and Gorbatikh, L., 2015. Matrix cracks around fibre breaks and their effect on stress redistribution and failure development in unidirectional composites. *Composites Science and Technology, 108*, pp. 16–22.
24. Kim, B.W. and Nairn, J.A., 2002. Observations of fiber fracture and interfacial debonding phenomena using the fragmentation test in single fiber composites. *Journal of Composite Materials, 36*(15), pp. 1825–1858.

25. Zhao, F.M. and Takeda, N., 2000. Effect of interfacial adhesion and statistical fiber strength on tensile strength of unidirectional glass fiber/epoxy composites. Part I: experiment results. *Composites Part A: Applied Science and Manufacturing, 31*(11), pp. 1203–1214.

26. Nedele, M.R. and Wisnom, M.R., 1994. Stress concentration factors around a broken fibre in a unidirectional carbon fibre-reinforced epoxy. *Composites, 25*(7), pp. 549–557.

27. Case, S.W., Carman, G.P., Lesko, J.J., Fajardo, A.B. and Reifsnider, K.L., 1995. Fiber fracture in unidirectional composites. *Journal of Composite Materials, 29*(2), pp. 208–228.

28. Asp, L.E., Berglund, L.A. and Talreja, R., 1996. A criterion for crack initiation in glassy polymers subjected to a composite-like stress state. *Composites Science and Technology, 56*(11), pp. 1291–1301.

29. Asp, L.E., Berglund, L.A. and Talreja, R., 1996. Prediction of matrix-initiated transverse failure in polymer composites. *Composites Science and Technology, 56*(9), pp. 1089–1097.

30. Asp, L.E., Berglund, L.A. and Gudmundson, P., 1995. Effects of a composite-like stress state on the fracture of epoxies. *Composites Science and Technology, 53*(1), pp. 27–37.

31. Neogi, A., Mitra, N. and Talreja, R., 2018. Cavitation in epoxies under composite-like stress states. *Composites Part A: Applied Science and Manufacturing, 106*, pp. 52–58.

32. Scott, A.E., Mavrogordato, M., Wright, P., Sinclair, I. and Spearing, S.M., 2011. In situ fibre fracture measurement in carbon–epoxy laminates using high resolution computed tomography. *Composites Science and Technology, 71*(12), pp. 1471–1477.

33. AhmadvashAghbash, S., Breite, C., Mehdikhani, M. and Swolfs, Y., 2022. Longitudinal debonding in unidirectional fibre-reinforced composites: Numerical analysis of the effect of interfacial properties. *Composites Science and Technology, 218*, 109117.

34. Yang, L. and Thomason, J.L., 2010. Interface strength in glass fibre–polypropylene measured using the fibre pull-out and microbond methods. *Composites Part A: Applied Science and Manufacturing, 41*(9), pp. 1077–1083.

35. He, M.-Y. and Hutchinson, J.W., 1989. Kinking of a crack out of an interface. *Journal of Applied Mechanics, 56*, pp. 270–278.

36. Swolfs, Y., Morton, H., Scott, A.E., Gorbatikh, L., Reed, P.A., Sinclair, I., Spearing, S.M. and Verpoest, I., 2015. Synchrotron radiation computed tomography for experimental validation of a tensile strength model for unidirectional fibre-reinforced composites. *Composites Part A: Applied Science and Manufacturing, 77*, pp. 106–113.

37. Thionnet, A., Chou, H.Y. and Bunsell, A., 2014. Fibre break processes in unidirectional composites. *Composites Part A: Applied Science and Manufacturing, 65*, pp. 148–160.

38. Graciani, E., Mantič, V., París, F. and Varna, J., 2009. Numerical analysis of debond propagation in the single fibre fragmentation test. *Composites Science and Technology, 69*(15–16), pp. 2514–2520.

39. Graciani, E., Varna, J., Mantič, V., Blázquez, A. and París, F., 2011. Evaluation of interfacial fracture toughness and friction coefficient in the single fiber fragmentation test. *Procedia Engineering, 10*, pp. 2478–2483.

40. Pupurs, A. and Varna J., 2013. FEM modeling of fiber/matrix debond growth in tension-tension cyclic loading of unidirectional composites. *International Journal of Damage Mechanics, 22*, pp. 1144–1160.

41. Pupurs, A. and Varna J., 2017. Steady–state energy release rate for fiber/matrix interface debond growth in unidirectional composites. *International Journal of Damage Mechanics, 26*, pp. 560–587.

42. Zhuang, L., Pupurs, A., Varna, J. and Ayadi, Z., 2016. Effect of fiber clustering on debond growth energy release rate in UD composites with hexagonal packing. *Engineering Fracture Mechanics, 161*, pp. 76–88.

43. Rosen, B.W., 1965. Mechanics of composites strengthening. In *Fiber Composite Materials*. American Society of Metals Seminar, ASM, Ohio, Chapter 3, pp. 37–75.
44. Budiansky, B., 1983. Micromechanics. *Computers and Structures*, *16*, pp. 1–4.
45. Argon, A.S. 1972. Fracture of composites. In H. Herman (Ed.),*Treatise of Materials Science and Technology*, Vol. 1. Academic Pres, New York, pp. 79–114.
46. Budiansky, B. and Fleck, N.A. 1993. Compressive failure of fibre composites. *Journal of the Mechanics and Physics of Solids*, *41*(1):183–211.
47. Wilhelmsson, D., Talreja, R., Gutkin, R. and Asp, L.E., 2019. Compressive strength assessment of fibre composites based on a defect severity model. *Composites Science and Technology*, *181*, 107685.
48. Wilhelmsson, D., Gutkin, R., Edgren, F. and Asp, L.E., 2018. An experimental study of fibre waviness and its effects on compressive properties of unidirectional NCF composites. *Composites Part A: Applied Science and Manufacturing*, *107*, pp. 665–674.
49. Wilhelmsson, D., Rikemanson, D., Bru, T. and Asp, L.E., 2020. Compressive strength assessment of a CFRP aero-engine component–an approach based on measured fibre misalignment angles. *Composite Structures*, *233*, 111632.
50. Hapke, J., Gehrig, F., Huber, N., Schulte, K. and Lilleodden, E.T., 2011. Compressive failure of UD-CFRP containing void defects: In situ SEM microanalysis. *Composites Science and Technology*, *71*(9), pp. 1242–1249.
51. Liebig, W.V., Viets, C., Schulte, K. and Fiedler, B., 2015. Influence of voids on the compressive failure behaviour of fibre-reinforced composites. *Composites Science and Technology*, *117*, pp. 225–233.
52. Liebig, W.V., Schulte, K. and Fiedler, B., 2016. Hierarchical analysis of the degradation of fibre-reinforced polymers under the presence of void imperfections. *Philosophical Transactions of the Royal Society A*, *374*, 20150279.
53. Hobbiebrunken, T., Hojo, M., Adachi, T., De Jong, C. and Fiedler, B., 2006. Evaluation of interfacial strength in CF/epoxies using FEM and in-situ experiments. *Composites Part A: Applied Science and Manufacturing*, *37*(12), pp. 2248–2256.
54. Elnekhaily, S.A. and Talreja, R., 2018. Damage initiation in unidirectional fiber composites with different degrees of nonuniform fiber distribution. *Composites Science and Technology*, *155*, pp. 22–32.
55. Elnekhaily, S.E., 2018. Damage initiation in unidirectional polymeric composites with manufacturing induced irregularities, PhD dissertation, Department of Materials Science and Engineering, Texas A&M University, College Station, Texas.
56. Sudhir, A. and Talreja, R., 2019. Simulation of manufacturing induced fiber clustering and matrix voids and their effect on transverse crack formation in unidirectional composites. *Composites Part A: Applied Science and Manufacturing*, *127*, 105620.
57. Dong, M. and Schmauder, S., 1996. Transverse mechanical behaviour of fiber reinforced composites—FE modelling with embedded cell models. *Computational Materials Science*, *5*(1–3), pp. 53–66.
58. Gamstedt, E.K. and Sjögren, B.A., 1999. Micromechanisms in tension-compression fatigue of composite laminates containing transverse plies. *Composites Science and Technology*, *59*(2), pp. 167–178.
59. Romanov, V.S., Lomov, S.V., Verpoest, I. and Gorbatikh, L., 2015. Modelling evidence of stress concentration mitigation at the micro-scale in polymer composites by the addition of carbon nanotubes. *Carbon*, *82*, pp. 184–194.
60. De Kok, J.M.M. and Meijer, H.E.H., 1999. Deformation, yield and fracture of unidirectional composites in transverse loading: 1. Influence of fibre volume fraction and test-temperature. *Composites Part A: Applied Science and Manufacturing*, *30*(7), pp. 905–916.
61. Toya, M., 1974. A crack along the interface of a circular inclusion embedded in an infinite solid. *Journal of the Mechanics and Physics of Solids*, *22*(5), pp. 325–348.

62. París, F., Cano, J.C. and Varna, J., 1996. The fiber-matrix interface crack—A numerical analysis using boundary elements. *International Journal of Fracture*, *82*(1), pp. 11–29.
63. Varna, J., Berglund, L.A. and Ericson, M.L., 1997. Transverse single-fibre test for interfacial debonding in composites: 2. Modelling. *Composites Part A: Applied Science and Manufacturing*, *28*(4), pp. 317–326.
64. García, I.G., Mantič, V. and Graciani, E., 2015. Debonding at the fibre–matrix interface under remote transverse tension. One debond or two symmetric debonds? *European Journal of Mechanics-A/Solids*, *53*, pp. 75–88.
65. París, F., Correa, E., Mantič, V., 2007. Kinking of transversal interface cracks between fiber and matrix. *Journal of Applied Mechanics*, *74* (4), pp. 703–716.
66. Sandino, C., Correa, E. and París, F., 2016. Numerical analysis of the influence of a nearby fibre on the interface crack growth in composites under transverse tensile load. *Engineering Fracture Mechanics*, *168*, pp. 58–75.
67. Zhuang, L., Talreja, R. and Varna, J., 2018. Transverse crack formation in unidirectional composites by linking of fibre/matrix debond cracks. *Composites Part A: Applied Science and Manufacturing*, *107*, pp. 294–303.
68. Zhuang, L., Pupurs, A., Varna, J., Talreja, R. and Ayadi, Z., 2018. Effects of interfiber spacing on fiber-matrix debond crack growth in unidirectional composites under transverse loading. *Composites Part A: Applied Science and Manufacturing*, *109*, pp. 463–471.
69. Sjögren, B.A. and Berglund, L.A., 2000. The effects of matrix and interface on damage in GRP cross-ply laminates. *Composites Science and Technology*, *60*(1), pp. 9–21.
70. Wood, C.A. and Bradley, W.L., 1997. Determination of the effect of seawater on the interfacial strength of an interlayer E-glass/graphite/epoxy composite by in situ observation of transverse cracking in an environmental SEM. *Composites Science and Technology*, *57*(8), pp. 1033–1043.
71. Sudhir, A. and Talreja, R., 2020. Corrigendum to "Simulation of manufacturing induced fiber clustering and matrix voids and their effect on transverse crack formation in unidirectional composites". [Compos. Part A: Appl. Sci. Manuf. 127 (2019) 105620]. *Composites Part A: Applied Science and Manufacturing*, *131*, 105705.
72. Elnekhaily, S.A. and Talreja, R., 2023. Effects of micro voids on the early stage of transverse crack formation in unidirectional composites. *Composites Part A: Applied Science and Manufacturing*, *167*, 107457.
73. Hojo, M., Mizuno, M., Hobbiebrunken, T., Adachi, T., Tanaka, M. and Ha, S.K., 2009. Effect of fiber array irregularities on microscopic interfacial normal stress states of transversely loaded UD-CFRP from viewpoint of failure initiation. *Composites Science and Technology*, *69*(11–12), pp. 1726–1734.
74. Sharifpour, F., Montesano, J. and Talreja, R., 2020. Assessing the effects of ply constraints on local stress states in cross-ply laminates containing manufacturing induced defects. *Composites Part B: Engineering*, *199*, 108227.
75. González, C. and LLorca, J., 2007. Mechanical behavior of unidirectional fiber-reinforced polymers under transverse compression: Microscopic mechanisms and modeling. *Composites Science and Technology*, *67*(13), pp. 2795–2806.
76. Correa, E., Mantič, V. and París, F., 2008. A micromechanical view of inter-fibre failure of composite materials under compression transverse to the fibres. *Composites Science and Technology*, *68*(9), pp. 2010–2021.
77. Correa, E., Mantič, V. and París, F., 2008. Numerical characterisation of the fibre–matrix interface crack growth in composites under transverse compression. *Engineering Fracture Mechanics*, *75*(14), pp. 4085–4103.

78. Sandino, C., Correa, E. and París, F., 2018. A study of the influence of a nearby fibre on the interface crack growth under transverse compression in composite materials. *Engineering Fracture Mechanics*, *193*, pp. 1–16.

79. Flores, M., Sharits, A., Wheeler, R., Sesar, N. and Mollenhauer, D., 2022. Experimental analysis of polymer matrix composite microstructures under transverse compression loading. *Composites Part A: Applied Science and Manufacturing*, *156*, 106859.

80. Elnekhaily, S.A. and Talreja, R., 2019. Effect of axial shear and transverse tension on early failure events in unidirectional polymer matrix composites. *Composites Part A: Applied Science and Manufacturing*, *119*, pp. 275–282.

81. Wu, X., Aramoon, A. and El-Awady, J.A., 2020. Hierarchical multiscale approach for modeling the deformation and failure of epoxy-based polymer matrix composites. *The Journal of Physical Chemistry B*, *124*(52), pp. 11928–11938.

82. Dugdale, D.S.,1960. Yielding in steel sheets containing slits. *Journal of the Mechanics and Physics of Solids*, *8*, pp. 100–104.

83. Barenblatt, G.I.,1962. The mathematical theory of equilibrium cracks in brittle fracture. In H.L. Dryden and T. Von Karman (eds.), *Advances in Applied Mechanics*. Academic Press, Elsevier Science, Amsterdam. pp. 55–129.

84. Rice, J.R., 1968. A path independent integral and the approximate analysis of strain concentration by notches and cracks. *Journal of Applied Mechanics, Transactions ASME*, *35*, 379–386.

85. Xu, X.-P. and Needleman, A., 1993. Void nucleation by inclusion debonding in a crystal matrix. *Modelling and Simulation in Materials Science and Engineering*, *1*, 111–132.

86. Segurado, J. and LLorca, J., 2005. A computational micromechanics study of the effect of interface decohesion on the mechanical behavior of composites. *Acta Materialia*, *53*(18), pp. 4931–4942.

87. Vaughan, T.J. and McCarthy, C.T., 2011. Micromechanical modelling of the transverse damage behaviour in fibre reinforced composites. *Composites Science and Technology*, *71*(3), pp. 388–396.

88. Vajari, D.A., Legarth, B.N. and Niordson, C.F., 2013. Micromechanical modeling of unidirectional composites with uneven interfacial strengths. *European Journal of Mechanics-A/Solids*, *42*, pp. 241–250.

89. Systémes, D., 2015. *Abaqus 6.14 Documentation–Theory Guide*. Providence.

90. Zhandarov, S. and Mäder, E., 2005. Characterization of fiber/matrix interface strength: Applicability of different tests, approaches and parameters. *Composites Science and Technology*, *65*(1), pp. 149–160.

91. García, I.G., Paggi, M. and Mantič, V., 2014. Fiber-size effects on the onset of fiber–matrix debonding under transverse tension: A comparison between cohesive zone and finite fracture mechanics models. *Engineering Fracture Mechanics*, *115*, pp. 96–110.

92. Zhang, C., Li, N., Wang, W., Binienda, W.K. and Fang, H., 2015. Progressive damage simulation of triaxially braided composite using a 3D meso-scale finite element model. *Composite Structures*, *125*, pp. 104–116.

93. Palizvan, M., Tahaye Abadi, M. and Sadr, M.H., 2020. Micromechanical damage behavior of fiber-reinforced composites under transverse loading including fiber-matrix debonding and matrix cracks. *International Journal of Fracture*, *226*, pp. 145–160.

94. Dimitri, R., Trullo, M., De Lorenzis, L. and Zavarise, G., 2015. Coupled cohesive zone models for mixed-mode fracture: A comparative study. *Engineering Fracture Mechanics*, *148*, pp. 145–179.

95. Bouhala, L., Makradi, A., Belouettar, S., Kiefer-Kamal, H. and Fréres, P., 2013. Modelling of failure in long fibres reinforced composites by X-FEM and cohesive zone model. *Composites Part B: Engineering*, *55*, pp. 352–361.

96. Kushch, V.I., Shmegera, S.V., Brøndsted, P. and Mishnaevsky Jr, L., 2011. Numerical simulation of progressive debonding in fiber reinforced composite under transverse loading. *International Journal of Engineering Science*, *49*(1), pp. 17–29.
97. Zhang, H., Ericson, M.L., Varna, J. and Berglund, L.A., 1997. Transverse single-fibre test for interfacial debonding in composites: 1. Experimental observations. *Composites Part A: Applied Science and Manufacturing*, *28*(4), pp. 309–315.
98. Li, S. and Ghosh, S., 2007. Modeling interfacial debonding and matrix cracking in fiber reinforced composites by the extended Voronoi cell FEM. *Finite Elements in Analysis and Design*, *43*(5), pp. 397–410.
99. Steinbach, I., 2009. Phase-field models in materials science. *Modelling and Simulation in Materials Science and Engineering*, *17*(7), 073001.
100. Wu, J.Y., Nguyen, V.P., Nguyen, C.T., Sutula, D., Sinaie, S. and Bordas, S.P., 2020. Phase-field modeling of fracture. *Advances in Applied Mechanics*, *53*, pp. 1–183.
101. Kristensen, P.K., Niordson, C.F. and Martínez-Pañeda, E., 2021. An assessment of phase field fracture: Crack initiation and growth. *Philosophical Transactions of the Royal Society A*, *379*(2203), 20210021.
102. Budiman, B.A., Takahashi, K., Inaba, K. and Kishimoto, K., 2016. Evaluation of interfacial strength between fiber and matrix based on cohesive zone modeling. *Composites Part A: Applied Science and Manufacturing*, *90*, pp. 211–217.
103. Sockalingam, S., Dey, M., Gillespie Jr, J.W. and Keefe, M., 2014. Finite element analysis of the microdroplet test method using cohesive zone model of the fiber/matrix interface. *Composites Part A: Applied Science and Manufacturing*, *56*, pp. 239–247.
104. Vajari, D.A., González, C., Llorca, J. and Legarth, B.N., 2014. A numerical study of the influence of microvoids in the transverse mechanical response of unidirectional composites. *Composites Science and Technology*, *97*, pp. 46–54.
105. Benzarti, K., Cangémi, L. and Dal Maso, F., 2001. Transverse properties of unidirectional glass/epoxy composites: Influence of fibre surface treatments. *Composites Part A: Applied Science and Manufacturing*, *32*(2), pp. 197–206.
106. Sharifpour, F., Montesano, J. and Talreja, R., 2022. Micromechanical assessment of local failure mechanisms and early-stage ply crack formation in cross-ply laminates. *Composites Science and Technology*, *220*, 109286.
107. Singh, C.V. and Talreja, R., 2010. Evolution of ply cracks in multidirectional composite laminates. *International Journal of Solids and Structures*, *47*(10), pp. 1338–1349.
108. Talreja R. and Singh, C.V., 2012. *Damage and Failure of Composite Materials*. Cambridge University Press.
109. Tong, J., Guild, F.J., Ogin, S.L. and Smith, P.A., 1997. On matrix crack growth in quasi-isotropic laminates—I. Experimental investigation. *Composites Science and Technology*, *57*(11), pp. 1527–1535.
110. Huang, Y., Varna, J. and Talreja, R., 2014. Statistical methodology for assessing manufacturing quality related to transverse cracking in cross ply laminates. *Composites Science and Technology*, *95*, pp. 100–106.
111. Sheinman, I. and Kardomateas, G.A., 1997. Energy release rate and stress intensity factors for delaminated composite laminates. *International Journal of Solids and Structures*, *34*(4), pp. 451–459.
112. Krueger, R. and O'Brien, T.K., 2001. A shell/3D modeling technique for the analysis of delaminated composite laminates. *Composites Part A: Applied Science and Manufacturing*, *32*(1), pp. 25–44.
113. Zou, Z., Reid, S.R., Soden, P.D. and Li, S., 2001. Mode separation of energy release rate for delamination in composite laminates using sublaminates. *International Journal of Solids and Structures*, *38*(15), pp. 2597–2613.

114. Tabiei, A. and Zhang, W., 2018. Composite laminate delamination simulation and experiment: A review of recent development. *Applied Mechanics Reviews*, 70(3), p. 030801.
115. Sørensen, B.F. and Jacobsen, T.K., 2003. Determination of cohesive laws by the J integral approach. *Engineering fracture mechanics*, 70(14), pp. 1841–1858.
116. Joki, R.K., Grytten, F., Hayman, B. and Sørensen, B.F., 2016. Determination of a cohesive law for delamination modelling–Accounting for variation in crack opening and stress state across the test specimen width. *Composites Science and Technology*, 128, pp. 49–57.
117. Benzeggagh, M.L. and Kenane, M.J.C.S., 1996. Measurement of mixed-mode delamination fracture toughness of unidirectional glass/epoxy composites with mixed-mode bending apparatus. *Composites Science and Technology*, 56(4), pp. 439–449.
118. Mathews, M.J. and Swanson, S.R., 2007. Characterization of the interlaminar fracture toughness of a laminated carbon/epoxy composite. *Composites Science and Technology*, 67(7–8), pp. 1489–1498.
119. Kim, B.W. and Mayer, A.H., 2003. Influence of fiber direction and mixed-mode ratio on delamination fracture toughness of carbon/epoxy laminates. *Composites Science and Technology*, 63(5), pp. 695–713.
120. Andersons, J. and König, M., 2004. Dependence of fracture toughness of composite laminates on interface ply orientations and delamination growth direction. *Composites Science and Technology*, 64(13–14), pp. 2139–2152.
121. Jordan, W.M., Bradley, W.L. and Moulton, R.J., 1989. Relating resin mechanical properties to composite delamination fracture toughness. *Journal of Composite Materials*, 23(9), pp. 923–943.
122. Albertsen, H., Ivens, J., Peters, P., Wevers, M. and Verpoest, I., 1995. Interlaminar fracture toughness of CFRP influenced by fibre surface treatment: Part 1. Experimental Results. *Composites Science and Technology*, 54, 133–145.
123. Asp, L. and Brandt, F., 1997, July. Effects of pores and voids on the interlaminar delamination toughness of a carbon/epoxy composite. In *Proceedings of 11th International Conference on Composite Materials*, Australia.
124. Olivier, P.A., Mascaro, B., Margueres, P. and Collombet, F., 2007, July. CFRP with voids: Ultrasonic characterization of localized porosity, acceptance criteria and mechanical characteristics. In *Proceedings of ICCM* (Vol. 16).
125. Abdelal, N.R., 2013. *Effects of Voids on Delamination Behavior Under Static and Fatigue Mode I and Mode II*. Dayton, OH: University of Dayton.
126. Ricotta, M., Quaresimin, M. and Talreja, R., 2008. Mode I strain energy release rate in composite laminates in the presence of voids. *Composites Science and Technology*, 68(13), pp. 2616–2623.
127. Short, G.J., Guild, F.J. and Pavier, M.J., 2001. The effect of delamination geometry on the compressive failure of composite laminates. *Composites Science and Technology*, 61(14), pp. 2075–2086.
128. Chai, H., Babcock, C.D. and Knauss, W.G., 1981. One dimensional modelling of failure in laminated plates by delamination buckling. *International Journal of Solids and Structures*, 17(11), pp. 1069–1083.
129. Kyoung, W.M. and Kim, C.G., 1995. Delamination buckling and growth of composite laminated plates with transverse shear deformation. *Journal of Composite Materials*, 29(15), pp. 2047–2068.
130. Whitcomb, J.D., 1981. Finite element analysis of instability related delamination growth. *Journal of Composite materials*, 15(5), pp. 403–426.
131. Lim, Y.B. and Parsons, I.D., 1993. The linearized buckling analysis of a composite beam with multiple delaminations. *International Journal of Solids and Structures*, 30(22), pp. 3085–3099.

132. Hwang, S.F. and Liu, G.H., 2001. Buckling behavior of composite laminates with multiple delaminations under uniaxial compression. *Composite Structures*, *53*(2), pp. 235–243.
133. Zhang, Y. and Wang, S., 2009. Buckling, post-buckling and delamination propagation in debonded composite laminates: Part 1: Theoretical development. *Composite Structures*, *88*(1), pp. 121–130.
134. Wang, S. and Zhang, Y., 2009. Buckling, post-buckling and delamination propagation in debonded composite laminates Part 2: Numerical applications. *Composite Structures*, *88*(1), pp. 131–146.
135. Zhuang, L. and Talreja, R., 2014. Effects of voids on postbuckling delamination growth in unidirectional composites. *International Journal of Solids and Structures*, *51*(5), pp. 936–944.
136. Jamison, R.D., 1985. The role of microdamage in tensile failure of graphite/epoxy laminates. *Composites Science and Technology*, *24*, pp. 83–99.
137. Hedgepeth, J.M. and van Dyke, P., 1967. Local stress concentrations in imperfect filamentary composite materials. *Journal of Composite Materials*, *1*, pp. 294–309.
138. Sun, C.T. and Tsai, J.L., 2001, June. Comparison of microbuckling model and kink band model in predicting compressive strength of composites. In *Proceedings of the 13th International Conference on Composite Materials*, Beijing, China.
139. Zhou, H.W., Yi, H.Y., Gui, L.L., Dai, G.M., Peng, R.D., Wang, H.W. and Mishnaevsky Jr, L., 2013. Compressive damage mechanism of GFRP composites under off-axis loading: Experimental and numerical investigations. *Composites Part B: Engineering*, *55*, pp. 119–127.
140. Shuart, M.J., 1989. Failure of compression-loaded multidirectional composite laminates. *AIAA Journal*, *27*(9), pp. 1274–1279.

9 Modeling of Fatigue Damage in the Presence of Defects

9.1 HISTORICAL DEVELOPMENT

Before discussing the modeling of fatigue damage, it will be useful to review the historical development of fatigue damage modeling for composite materials. In the early years of fatigue damage modeling for composites, i.e., in the late 1960s to mid-1970s, it was common to use the concepts developed for metal fatigue. The literature often referred to damage "initiation" and damage "growth", analogous to crack initiation and crack growth in metals. Fatigue failure was described as the degradation of residual strength to the level of the maximum cyclic load applied. Figure 9.1 depicts the concept where the crack size increase with the time to service of a component is also shown. The relationship between the crack size and the residual strength, shown in the figure, was given by the fracture mechanics-based failure criterion that required knowing the fracture toughness of the material.

Later, in the late-1970s, observations were reported that showed a multitude of cracks developing under cyclic loading of composite laminates and failure did not result from a critical size of a single crack. Based on a host of studies, a fatigue damage development schematic was presented, as illustrated in Figure 9.2 [1]. Alongside these observations, the measurement of average elastic modulus in the loading direction showed variation with cycles, as illustrated in Figure 9.3 [1].

These and similar observations prompted the field of continuum damage mechanics (CDM) that developed thermodynamics-based internal variable formulations to relate the crack density to stiffness coefficients of composite laminates [2–7]. It would be useful to note at this point that the experimental data such as that shown in Figure 9.3 correlated with a specific measured variable, which is the number of cracks per unit axial length of the specimen, labeled as "transverse crack density" in Figure 9.3. Unfortunately, it became common in the literature to show "stiffness degradation" as a function of "damage" for the entire damage development until failure without quantifying the damage beyond the crack density, which was measured only in the early stage of damage development, ending in a state of constant crack density, labeled as CDS (characteristic damage state) in Figure 9.2.

The theoretical underpinning in CDM was indeed only for distributed internal entities that could be carried as damage variables in the formulations [2–7]. Thus, for stages of damage past the CDS in Figure 9.2 there was no justification to relate

DOI: 10.1201/9781003225737-9

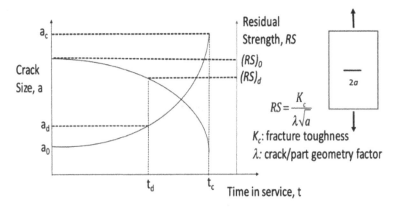

FIGURE 9.1   A crack growth-based residual strength degradation with time in service.

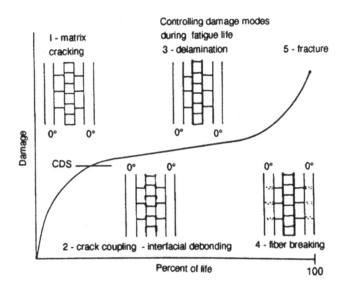

FIGURE 9.2   A schematic representation of fatigue damage development in composite laminates. Source: Jamison, R. D., Schulte, K., Reifsnider, K. L., & Stinchcomb, W. W., 1984.
Characterization and analysis of damage mechanisms in tension-tension fatigue of graphite/
epoxy laminates. *Effects of Defects in Composite Materials, ASTM STP, 836*, 21–55.

stiffness (e.g., secant modulus measured by an extensometer over a gauge length
of typically 25 mm) to the same damage measure as pre-CDS. Other CDM formulations, such as [8–10] did not use observable internal changes in a material, e.g.,
matrix cracks, fiber–matrix debonds, and interior delaminations, all of which have
the nature of being distributed, in defining damage variables. Instead, they cast
the damage measures in terms of the consequence of damage, e.g., the average (or
effective) property changes resulting from those internal changes. This led to the

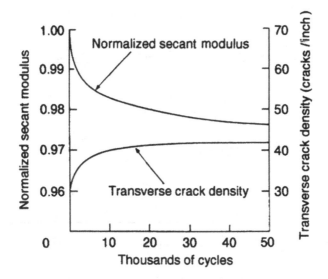

FIGURE 9.3   Correspondence of the secant modulus, normalized by its initial value, and the number density of transverse cracks in a cross-ply laminate. Source: Jamison, R. D., Schulte, K., Reifsnider, K. L., & Stinchcomb, W. W., 1984. Characterization and analysis of damage mechanisms in tension-tension fatigue of graphite/epoxy laminates. *Effects of Defects in Composite Materials, ASTM STP, 836,* 21–55.

detachment of the physical damage mechanisms from the materials' response characteristics and has to this day contributed to the misunderstanding of the damage concept pertaining to CDM. The rich concept of damage was reduced to a number varying from 0 to a critical value, often 1. A more recent effort [11] introduced multiple damage variables, each associated with a selected "orientation" of damage such as longitudinal and transverse with respect to the fiber direction in a ply of a laminate. Still, no physical measures of the damage entities themselves (e.g., their densities) were introduced, and their values were extracted from measured changes in the stiffness coefficients.

Explicit incorporation of distributed physical damage entities (matrix cracks, fiber–matrix debond cracks, interface cracks, etc.) in CDM allowed connecting the stiffness property changes formulated by it to specific micromechanics solutions based on crack surface displacements [12–16]. In fact, the constants appearing in the CDM formulation could be interpreted by using micromechanics solutions [17].

Returning to strength degradation as another approach to describing damage development in fatigue, it was postulated that although no criticality related to the size of a single crack existed in composite laminates, a critical element can trigger final failure [18]. Such a critical element was envisioned as a localized zone formed by interconnected cracks at ply interfaces. Failure from this element was thought to come from its local driving forces supplied by the subcritical elements, i.e., the failure events such as distributed matrix cracks. Figure 9.4 depicts the fatigue life locus where the strength of the critical element degrades to the applied maximum

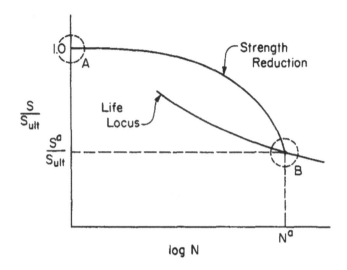

FIGURE 9.4 Schematic illustration of the fatigue life locus described by the condition of the initial strength at A reducing to the level of the maximum cyclic load S^a at B. S_{ult} is the ultimate tensile load. Source: Reifsnider, K.L., 1986. The critical element model: a modeling philosophy. *Engineering Fracture Mechanics*, 25(5–6), 739–749.

load [18]. An analogy depicted by this figure to the single crack-based service life using the fracture mechanics-based residual strength (Figure 9.1) is clear. While the crack size can be measured, it is not clear how the critical element size can be determined.

To summarize, the approaches to fatigue life prediction, Figure 9.5, describes an overall methodology. To begin, one conducts stress analysis in critical sites ("hot spots") of a composite structure using the deformation models (elasticity, viscoelasticity, plasticity, etc.) for initial (undamaged, but containing manufacturing defects) material state under specified service loading (mechanical, thermal, moisture, etc.). The damage mechanisms operating under the local stress states at the critical sites are identified based on experimental studies. Damage mechanics (combined CDM and micromechanics) analyses are applied at multiple scales to predict the occurrence of the first failure event ("initiation") and sequential subsequential failure events ("progression") culminating in final failure (material separation). Two alternative paths to life prediction then involve design performance-related criticalities. The strength-based life prediction concerns the loss of load-bearing capacity, while the stiffness-based approach takes account of the fact that changes in deformation characteristics (e.g., vibrational modes) may cause functional failure, although there may still be a load-carrying capacity. The stiffness degradation path should update the initial deformation models accounting for damage-induced changes.

In the following, fatigue life prediction of UD composites will be considered first, analyzing the damage development against the backdrop of FLDs described

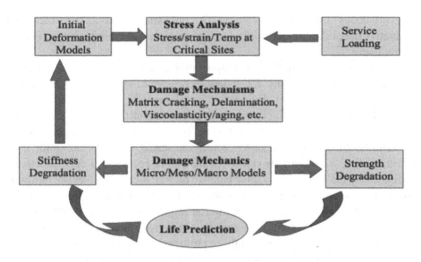

FIGURE 9.5 A flowchart for damage mechanics-based life prediction methodology.

in Chapter 7. Laminates will be considered next where multiaxial loading and inter-laminar damage will be given appropriate attention. Considerations of manufacturing defects will be made following the modeling approaches.

9.2 UD COMPOSITES

9.2.1 LONGITUDINAL TENSION–TENSION FATIGUE

In many applications where a cyclic uniaxial tensile load is of main concern in design, plies are placed in laminates oriented in the load direction. In such cases, modeling of UD composites under longitudinal loading is critical. Since fibers share most of the load in practical levels of fiber volume fraction (>50%), it is important to know how their contribution to the load-carrying capacity is affected by repeated tension. This problem is conceptually different from metal fatigue in cyclic tension where crack initiation, stable crack growth, and unstable crack growth are the three stages of the fatigue process. The crack initiation stage can be ignored if a stress concentration site such as a sharp defect or a notch or cutout exists. In that case, the focus is on the growth of a single crack for which the field of fracture mechanics has provided the basis for addressing that scenario. In UD composites, however, discrete fibers embedded in a polymer matrix create a complex scenario where the response to repeated loading is different in fibers and in the matrix. After much discussion in the early years of composite fatigue studies it was considered settled that if there is any fatigue failure in glass and carbon fibers, it is negligible in comparison to the well-defined fatigue damage development in polymers. In view of the highly statistical nature of failure under quasi-static tension in fibers, it seems an unnecessary complication to introduce further uncertainty in modeling by considering the fatigue of fibers. As will be discussed in the sequel, understanding the role of (instantaneous) fiber failures within UD composites where matrix cracking and fiber–matrix debonding under cyclic loading are plausible failure events, is a challenge by itself.

The process of tension–tension fatigue in UD composites was schematically described in Chapter 7, Figure 7.2, where the essential statistical nature of failure progression induced by fiber failures was explained. The experimental evidence gathered in [19] supported the conceptual framework of three regions in FLDs for carbon/epoxy and carbon/PEEK composites. To recall, in Region I, where the maximum strain in the first applied load is within the scatter band of quasi-static failure, the failure progression is highly statistical and can be regarded as cycle-independent (highlighted by the horizontal scatter band). The statistical nature of the failure progression is moderated, however, in Region II where gradual (not instantaneous) failure in the matrix and at fiber–matrix interfaces provides a cycle-dependent rate of progression. The fiber failure in Region II is still instantaneous (i.e., of negligible cycle dependence). Thus, the cyclic failure progression can be regarded as controlled by matrix and fiber–matrix debonding with fiber failures influencing the rate of progression, more at the upper end of Region II than as Region III is approached where the dividing threshold (fatigue limit) is given by insufficient progression to cause failure in a specified large number of cycles.

As described in Chapter 7, Section 7.2.1, ten years after the fatigue damage mechanisms study for a carbon/epoxy composite was reported [19], a study for glass/epoxy was performed [20]. That study focused on the role of fibers in the fatigue damage process. As was pointed out in the first FLD paper by this author [21], the extent of Region I in the FLD for glass/epoxy composite is small (a few hundred cycles) and is therefore easy to miss in examining the role of fibers in this region. The observations made in [20] were in Region II and inferences drawn there were concerning fatigue damage development and the fatigue limit. One of the main issues the authors in [20] set out to resolve was whether the final failure in fatigue comes from a cluster of broken fibers, as suggested by observations made in quasi-static loading [23, 24], or does it come from a fiber-bridged matrix crack as postulated in [22] and inferred from observations in [19], as described in Chapter 7.

The issue raised in the study [20] could not be fully resolved as the observations on fiber–matrix debonding were not quantified because the debond cracks tended to close on unloading of the specimen used for SEM observations. The main conclusion drawn was that fiber failures in the first application of the load were the necessary mechanism for fatigue damage progression. Logically, therefore, the fatigue limit will be given by the threshold for no fiber failures. One can argue that assuming no fiber failures in the first load application, irreversibility of deformation is still possible in the matrix (in particular, in resin-rich zones), and an eventual fatigue crack in the matrix can trigger fiber failures, which can activate the formation of a critical cluster of broken fibers. A thought experiment with composites of different fiber volume fractions subjected to tension–tension fatigue suggests that as the fiber volume fraction approaches zero (unreinforced composite), the fatigue process will tend to that of the polymer. This was the basis for suggesting that the lowest possible fatigue limit of UD composites will be given by an experimentally obtained fatigue limit of the polymer under cyclic strain [21, 22]. This limit will be enhanced as the fibers are made stiffer and their volume fraction is increased.

The role of fiber–matrix debonding as the mechanism of irreversibility, and therefore contributing to the progression of fatigue damage, must be considered in modeling fatigue damage. In [19] the debonding was studied in Region III of FLD for carbon/epoxy as a mechanism for slowing down and arresting the growth of a fiber-bridged matrix crack. This provided a plausible explanation for the occurrence of the fatigue limit. The role of debonding in the upper end of Region II was, however, not clarified. The crack length in this regime is expected to be small and its modeling by a fiber-bridged crack may not be effective. Instead, the starting point could be to model quasi-static failure from unstable growth of a "fracture plane", as proposed in [25]. Observations reported in [26] on a carbon/epoxy UD composite by SEM provide evidence of failure induced in an intact fiber by a debond crack emanating from a broken fiber end.

Guided by the fatigue damage mechanisms described above, a fatigue life prediction model was reported in [27, 28]. The model combined two essential features of the fatigue damage: a) the statistical nature of fiber failures, and b) the cycle-dependent growth of fiber–matrix debonding. The statistical fiber failure was described by the weak link theory and multiple fiber failures were treated by the so-called hierarchical scaling model [29]. The initiation of fiber–matrix debonding was modeled with a linear cohesive law defined by the shear strength and mode II fracture toughness. A Paris law was assigned to the self-similar debond crack growth.

The fatigue life prediction of the model [27, 28] is shown in Figure 9.7 with the experimental data from [19] plotted in the figure. The main support of the model is

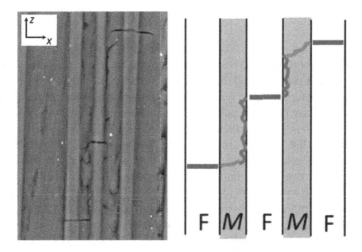

FIGURE 9.6 A failure scenario showing the progression of fiber breaks aided by matrix cracking and fiber–matrix debonding [26]. Source: Pagano, F., 2019. *Mécanismes de fatigue dominés par les fibres dans les composites stratifiés d'unidirectionnels* (Doctoral dissertation, Paris Sciences et Lettres (ComUE)).

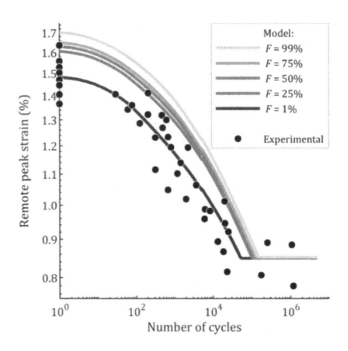

FIGURE 9.7   Prediction of fatigue life of a carbon/epoxy UD composite by the approach in Pimenta, S., Mersch, A. and Alves, M., 2018, July. Predicting damage accumulation and fatigue life of UD composites under longitudinal tension. In *IOP Conference Series: Materials Science and Engineering* (Vol. 388, No. 1, 012007). IOP Publishing, with experimental data from: Gamstedt, E.K. and Talreja, R., 1999. Fatigue damage mechanisms in unidirectional carbon-fibre-reinforced plastics. *Journal of Materials Science, 34,* 2535–2546.

in the prediction of the characteristic features of the FLD, i.e., the presence of three distinct regions. It is recognized that a precise prediction of fatigue life by the model would require determining several material parameters from independent tests, which is not an easy task. The authors conducted a sensitivity study of some of the parameters to gain insight into their contribution to the fatigue process. For instance, the sensitivity to the Weibull distribution of fiber strength is shown to be high in Region I, which is essentially governed by the static strength (idealized in FLD by a horizontal scatter band), see Figure 9.8. The high sensitivity to the fiber strength is seen to reduce as transition is made to Region II of FLD where the cycle dependency comes largely from the fatigue growth of the fiber–matrix debond cracks. The sensitivity of the fatigue limit to the fiber–matrix interface crack growth is illustrated in Figure 9.9 where the fatigue limit for three different values of the threshold of the energy release rate is shown. The zero value of this parameter corresponds to no debond initiation.

A different study reported in [30] assumed a "damage zone" formed by interactive fiber failures as the basis for analyzing progressive fatigue mechanisms.

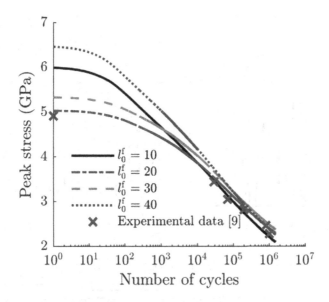

FIGURE 9.8 The fatigue life prediction as affected by the Weibull distribution parameters measured on different fiber gauge lengths. Source: Alves, M. and Pimenta, S., 2018. A computationally-efficient micromechanical model for the fatigue life of unidirectional composites under tension-tension loading. *International Journal of Fatigue*, *116*, 677–690.

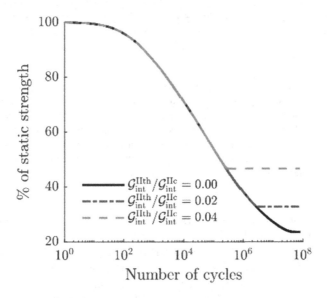

FIGURE 9.9 The sensitivity of the fatigue limit to the energy release rate threshold for debonding crack growth in mode II. Three values of the parameter are indicated. Source: Pimenta, S., Mersch, A. and Alves, M., 2018, July. Predicting damage accumulation and fatigue life of UD composites under longitudinal tension. In *IOP Conference Series: Materials Science and Engineering* (Vol. 388, No. 1, 012007). IOP Publishing.

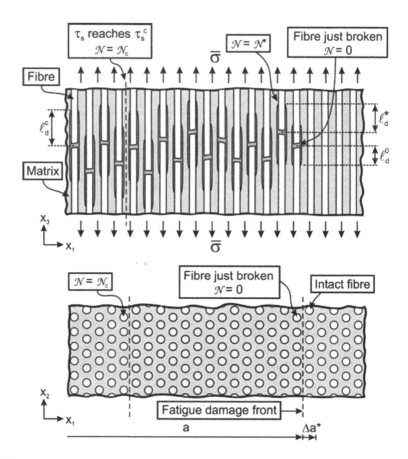

FIGURE 9.10 Illustration of a fatigue damage zone showing in the longitudinal view (top) broken fibers and debonding emanating from the broken fiber ends. The cross-sectional view (bottom) shows the size of the damage zone a, and its advancement Δa^*, caused by breakage of an intact fiber. The interfacial shear stress τ and the corresponding debond length l_d are shown. Their initial, current, and critical values are indicated by superscripts ^{0}, *, and ^{c}, respectively. Source: Sørensen,B.F., S.Goutianos. 2019. Micromechanical model for prediction of the fatigue limit for unidirectional fibre composites. *Mechanics of Materials*, 131, 169–187.

Longitudinal and cross-sectional views of the damage zone are depicted in Figure 9.10. As illustrated in the figure, each fiber break is accompanied by fiber–matrix debonding and interfacial shear stress acting over the debond length. The advancement of the damage zone length a in the projected longitudinal plane is assumed to occur by the failure of the next and subsequent fibers. The cyclic growth rate of the debond length da/dN provides the basis for predicting fatigue life with a failure criterion in terms of the critical damage zone length. It is noted that the damage front is taken as straight, and its advancement is assumed self-similar to allow treating it like a crack of length a.

The interaction between a fiber just broken and a neighboring intact fiber, i.e., stress enhancement on an intact fiber caused by the broken fiber, is assumed to occur via the intense stress field carried at the fiber–matrix interface debond crack front. The shear stress driving the debond crack in mode II is assumed to be "frictional", i.e., the two surfaces of the debond crack are assumed to be in frictional contact (presumably because of asperities on the crack surfaces). The frictional shear stress is assumed to decrease with load cycles, e.g., by an abrasive wear of the asperities, leading to reduced resistance to the debond crack growth and hence increased debond length.

The progressive fiber failure process is illustrated in Figure 9.11 for three fibers [30]. Beginning with the breakage of the first (weakest) fiber (Fiber #1), an instantaneous debonding of the fiber–matrix interface over a certain length occurs. The (frictional) shear stress developed over this debond length reduces with cycles, increasing the debond length and thereby increasing the length of the neighboring intact fibers over which enhanced axial stress occurs. At a certain number of cycles, a neighboring intact fiber can break if its strength over the affected length is exceeded by the enhanced stress. This process then continues to the breakage of the next fiber. The cluster of broken fibers thus formed contains broken fibers distributed in the cross-sectional plane and staggered in the longitudinal plane, as illustrated in Figure 9.11 d. The staggered pattern of fiber breaks agrees with the observed one shown in Figure 9.6.

There are several parameters in the modeling methodology proposed in [30] that need to be evaluated or estimated. Other than the elastic moduli of the constituents (only Young's moduli as the analysis is one-dimensional), failure-related material properties are needed. For fibers, the failure is described by a two-parameter Weibull distribution in which the characteristic strength and the exponent (Weibull modulus) are needed. Matrix failure does not enter the model. The fiber–matrix interface is characterized by the frictional shear stress and the critical energy release rate of a bi-material crack tip, which is simplified to its mode II component. The cycle-dependent decay of the interfacial shear stress is assumed as a power law in which the difference of the initial and final values of the shear stress and a decay exponent enter. These parameters need to be estimated. Finally, the growth rate of the damage zone da/dN, assumed as a Paris law in which a multiplying constant and an exponent are entered, needs to be extracted from the data. The critical damage zone size at failure is assumed in terms of the longest debond length achieved. This length is estimated from the mode II fracture toughness of the composite, which in turn is estimated from the matrix and interface mode II fracture toughness values. This adds knowing the mode II fracture toughness of the matrix to the list of model parameters.

The authors of the study [30] conducted parametric studies to gain insight into the sensitivity of the model predictions to the parameters. Figure 9.12 illustrates how the damage zone growth rate varies with the applied cyclic peak strain depending on the decay exponent n in the assumed frictional shear stress power law. While the sensitivity of the growth rate to the decay exponent is high, a notable feature is the insensitivity of the lowest growth rate to the decay exponent. The applied maximum strain in the first cycle corresponding to this growth rate is interpreted as the strain

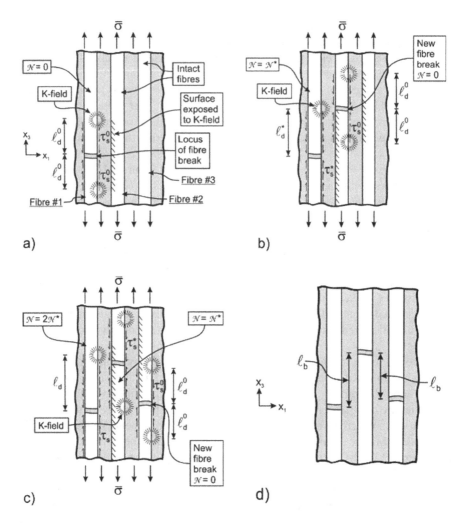

FIGURE 9.11 Illustration of the formation of a three broken-fiber cluster. a) The first broken fiber (Fiber #1) debonds instantaneously over an initial debond length. The intense stress field (K-field) carried by the debond crack front enhances the axial stress in Fiber #2. This length increases as the interfacial shear stress decreases with cycles. b) Fiber #2 breaks at a certain number of cycles, and the process continues similarly to c) when breakage of Fiber #3 occurs. d) The final staggered pattern of the three fiber breaks. Source: Sørensen,B.F., S.Goutianos. 2019. Micromechanical model for prediction of the fatigue limit for unidirectional fibre composites. *Mechanics of Materials*, 131, 169–187.

whose repetition does not produce new fiber failures. Thus, this value can be taken as the fatigue limit. The authors calculate this estimate of the fatigue limit to depend inversely on the fiber volume fraction.

The fatigue life prediction of the model is depicted in Figure 9.13 for a glass/epoxy composite where experimental data [31] are also shown. With the appropriate

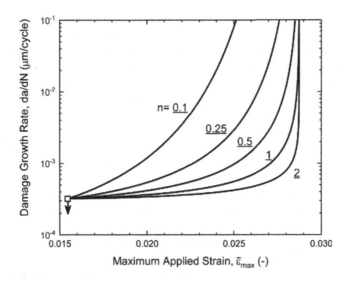

FIGURE 9.12 The growth rate of the damage zone length a calculated by the model [30] for different values of the decay exponent in the interfacial shear stress power law. Source: Sørensen,B.F., S.Goutianos. 2019. Micromechanical model for prediction of the fatigue limit for unidirectional fibre composites. *Mechanics of Materials*, 131, 169–187.

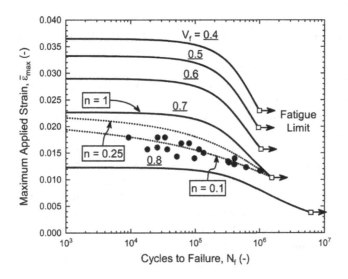

FIGURE 9.13 Fatigue life prediction of the model for a glass/epoxy UD composite. The parameters n and V_f are the frictional shear stress decay exponent and the fiber volume fraction, respectively. Source: Sørensen,B.F., S.Goutianos. 2019. Micromechanical model for prediction of the fatigue limit for unidirectional fibre composites. *Mechanics of Materials*, 131, 169–187, with experimental data from Korkiakoski, S., Brøndsted, P., Sarlin, E. and Saarela, O., 2016. Influence of specimen type and reinforcement on measured tension–tension fatigue life of unidirectional GFRP laminates. *International Journal of Fatigue*, 85, pp.114–129.

choice of the interfacial shear stress decay exponent, the prediction agrees well with the data [30]. However, the fiber volume fraction also needed adjustment to get a good fit to the data. Clearly, the average fiber volume fraction does not represent the local conditions in a composite that govern the interactive process of fiber failures. Also, the key source of damage accumulation assumed in the model relies on the frictional shear stress at the fiber–matrix interface. Not only is friction a difficult phenomenon to model, but its decay with cycles is also not fully understood.

The two modeling strategies described above have some common features while differing in detail. First, the statistical nature of fiber failures is accounted for by a weak link theory in both approaches. Consequently, the sequential progression of fiber failures is recognized in both approaches as causing randomness in progression and hence in the fatigue life. The irreversibility underpinning damage accumulation is, however, approached differently in the two approaches. In [27], the cycle-induced accumulation is attributed to the growth of the fiber–matrix debond crack driven by the mode II energy release rate. A Paris-type power law is assumed to describe the self-similar growth of the crack. In [30], the surfaces of the debond crack are assumed to be in frictional contact, driven by the frictional shear stress at the interface. A power law is assumed for the decay with cycles of this stress, leading to the progression of the debond length. Both models need interfacial fracture toughness in mode II but differ in characteristic values of the interfacial shear stress needed due to the way this stress is viewed: Without friction in [27], and with friction in [30].

It is noted that both modeling approaches do not account for matrix failure. The interaction between fibers responsible for an intact fiber failing in the vicinity of a broken fiber is treated entirely via fiber–matrix debonding. This is physically justifiable when fibers are closely spaced. In a more general case, the role of matrix failure may not be ignorable. In fact, for large separation distances between fibers, one can get insight into the relative roles of matrix cracking and fiber–matrix debonding from experimental observations on single-fiber fragmentation tests. Based on observations reported on cracks emanating from a broken fiber end [32, 33], the cracking modes were summarized in [34] and are shown in Figure 9.14. Interfacial debonding is seen often when the fiber–matrix interface is expected to be weak, e.g., when the sizing on the fiber surface is either inefficient or absent. If, on the other hand, the fiber–matrix bonding is controlled to be strong (good sizing, full wetting of fiber surface with polymer, proper curing of polymer, etc.), then a crack emanating from the broken fiber end extends into the matrix at different angles depending on the local stress state. The most likely scenario is a crack inclined nearly parallel to the fiber and growing away from the fiber direction. An analysis reported in [34] predicted the path of a matrix crack in a single-fiber composite and in [25] for multi-fiber composites.

9.2.1.1 Effect of Manufacturing Defects

The two approaches to modeling the fatigue process described above have the potential to account for manufacturing defects. In fact, using a weak link theory to describe the statistical strength of fibers indirectly accounts for the surface defects in

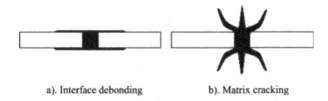

a). Interface debonding   b). Matrix cracking

FIGURE 9.14 Modes of cracking emanating from a broken fiber end, a) interfacial debonding, and b) matrix cracking. Source: Huang, H. and Talreja, R., 2006. Numerical simulation of matrix micro-cracking in short fiber reinforced polymer composites: initiation and propagation. *Composites Science and Technology, 66*(15), 2743–2757.

fibers that are the cause of failure. A non-uniform distribution of fibers in the cross section of a UD composite caused by manufacturing will produce fiber clusters and resin-rich zones. The modeling efforts described above do not consider the presence of resin-rich zones, which are likely to initiate matrix cracks. Such cracks can be arrested by stiff fibers, or the fibers can break, allowing the matrix crack to advance. This scenario is likely to develop towards the lower end of Region II of FLD and could be a cause of the fatigue limit when the growth rate of the fiber-bridged crack is too low to reach the criticality needed for composite failure.

Matrix voids resulting from manufacturing can conceivably affect the fatigue performance of UD composites in tension–tension cycling. However, controlled tests to investigate this effect have not been done. It is possible that a void along a fiber locally influences the fiber–matrix interface bonding resulting in overstressing of the fiber in tension. This could trigger fiber failure if the local stress in the fiber exceeds its strength. The presence of a void can also initiate a fatigue crack in the matrix by local stress concentration. On the other hand, a void in the path of a growing matrix crack can also blunt the crack tip and slow down its growth rate.

9.2.2 LONGITUDINAL COMPRESSION–COMPRESSION FATIGUE

Under cyclic loading in the fiber direction, load excursions in tension versus in compression have fundamentally different effects on the fatigue process. The tension case was considered in constructing the baseline FLD. The variation to this diagram due to compression loading will be discussed next.

For the case of compression along fibers in a UD composite, fiber microbuckling is a possible failure mechanism [35]. Experimental observations have, however, indicated that the fibers tend to buckle under the additional influence of local shear at stress concentration sites such as defects in the matrix and misaligned or wavy fibers. If totally elastic (reversible) conditions persist surrounding fiber microbuckling, then no further changes can be expected under repeated load excursions in compression. However, if the polymer matrix deforms inelastically in the region of a microbuckled fiber, then this irreversible deformation process will intensify the fiber buckling, thereby inducing microbuckling of the neighboring fibers, leading eventually to the formation of a kink band. This progressive failure mechanism is illustrated in

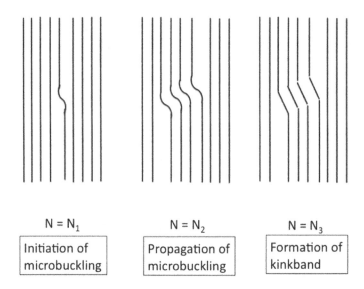

$N = N_1$ | $N = N_2$ | $N = N_3$

| Initiation of microbuckling | Propagation of microbuckling | Formation of kinkband |

FIGURE 9.15 Schematic illustration of compressive failure mechanism in a UD composite.

Figure 9.15, highlighting essential features without secondary aspects such as the broadening of kink bands during their progression. Final failure (often described as shear crippling) in fatigue is expected when sufficient fibers break at the kinks in the kink bands, leading to loss of the load-carrying capacity.

As described above, the failure process in UD composites under cyclic compression has phases of initiation, progression, and criticality. If fiber microbuckling occurs under the first application of a compressive load accompanied by inelastic deformation of the matrix, then in subsequent load applications, progression of microbuckling, kink-band formation, and fiber breakage are possible. The rates at which the post-microbuckling mechanisms occur will depend on the maximum level and range of the repeating stress. If the maximum stress level is within the scatter band of the compressive strength, and the composite survives, then failure will be expected in a few subsequent load applications. Thus, conditions for non-progressive fiber failure that characterize Region I in the baseline FLD will not be present in compression fatigue. In other words, Region II of the FLD (corresponding to progressive failure in compression) will start from the static strength and will slope downwards with cycles. A continuous slope in Region II will be expected if the nature of progressive failure mechanisms does not change with the load level. Fatigue limit will then be given by the threshold stress below which no inelastic fiber microbuckling occurs in the first application of load. This stress level is expected not only to depend on the fiber properties (e.g., stiffness) but also on the presence of manufacturing-induced defects such as fiber misalignment and waviness.

An attempt to model fatigue life in cyclic longitudinal compression was made in [36] where the starting point was the Argon model for kink-band formation discussed in Chapter 8, Section 8.4.2. Accordingly, the axial compressive stress σ_c for kink-band formation is given by [37],

$$\sigma_c = \frac{\tau_y}{\phi} \tag{9.1}$$

where τ_y is the shear yield strength of the UD composite. The authors in [36] conjectured that this strength value degrades under applied cyclic compression due to degradation of the matrix caused by fatigue microcracks formed in the matrix. Consequently, when the applied axial stress reaches the reduced σ_c for incipient kink-band formation, failure initiates. Clearly, failure will initiate at the site where the largest misalignment of angle ϕ exists. Assuming a power law for the degradation rate per cycle of the shear yield strength, the authors derived an expression for the fatigue life where the unknown value of the misalignment angle ϕ appears. This value can in principle be estimated following the methods described in Chapter 3, Section 3.3, and in Chapter 8, Section 8.4.2.1.

9.2.2.1 Effect of Voids

As stated before, controlled experiments to observe the effect of voids in aiding compressive failure are difficult to perform. One study [38] prepared a model UD glass/epoxy composite containing eight E-glass fibers of 70 mm in diameter. An air bubble was placed between fibers to study the void influence on compressive failure initiation and progression. Based on photoelasticity observations the authors summarized the failure process depicted in Figure 9.16. Although these observations are under quasi-static loading, they provide insight into the role of a void in locally degrading the shear response of the composite under repeated loading. Thus, they confirm the conjecture made in [36] regarding the shear yield strength reduction of the composite under cyclic compression.

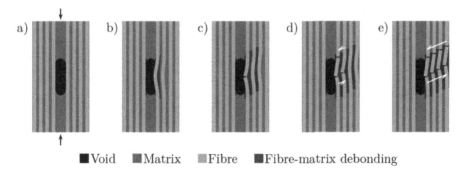

■Void ■Matrix ■Fibre ■Fibre-matrix debonding

FIGURE 9.16 Illustration of the compressive failure process in the presence of a void. a) Beginning of the compression load, b) buckling of a fiber next to the void, c) failure of the buckled fiber, d) shear band formation by sequential fiber breaks, and e) kink-band formation. Source: Liebig, W.V., Viets, C., Schulte, K. and Fiedler, B., 2015. Influence of voids on the compressive failure behaviour of fibre-reinforced composites. *Composites Science and Technology, 117,* 225–233.

9.2.3 COMBINED CYCLIC LOADING: MULTIAXIAL FATIGUE

Fatigue testing under combined cyclic loading is a challenging field. The historical development of this field and the progress made have been reviewed in [39]. Keeping our focus on the effects of manufacturing defects, only those studies that clarify the nature of failure events under repeated loads are of interest. In [40] test data gathered by testing different specimen types, e.g., off-axis flat specimens and tubular specimens under axial tension and torsion, were reviewed. In view of the practical limitations in imposing combined cyclic loading on specimens, it is efficient to focus on damage in the matrix and at the fiber–matrix interface separately from fiber failures in selecting combinations of loading. Thus, on-axis loading that causes fiber-dominated failure is complemented by off-axis loading at different inclinations to promote damage in the inter-fiber region. If the in-plane shear induces failure in the matrix and if this failure is not affected by tension in fibers as long as they remain intact, one can ask: How does damage accumulate in the matrix under cyclic shear when fibers are present? The observations related to static failure in in-plane shear were described in Chapter 5, Section 5.6 and the related modeling was discussed in Chapter 8, Section 8.5. Having this understanding, we need to extend it to the accumulation of inelastic shear strain with load cycles. Similarly, the initiation and progression of failure of the fiber–matrix interface need to be understood. In describing the effect of matrix fatigue damage with intact fibers on the baseline FLD it was contemplated that the accumulated inelastic strain in the matrix will produce microcracks that will eventually coalesce and grow unstably along the fibers [22]. Figure 9.17 taken from that publication depicts the assumed failure process. As illustrated in the figure, the baseline FLD (cyclic tension, $\theta = 0$) loses the fiber failure dominated

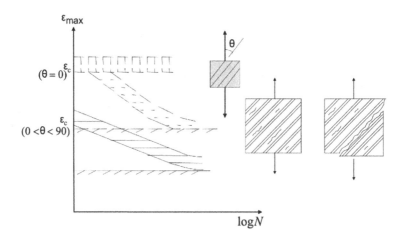

FIGURE 9.17 Changes in the baseline FLD with the off-axis angle q [22]. Source: Talreja, R., 1981. Fatigue of composite materials: damage mechanisms and fatigue-life diagrams. *Proceedings of the Royal Society of London. A. Mathematical and Physical Sciences*, 378(1775), 461–475.

Region I and Region II inclines increasingly with the off-axis angle θ of the cyclic tension direction. As this angle approaches 90°, the failure becomes dominated by fiber–matrix interfacial failure (also known as transverse fiber debonding). Figure 9.18 provides support for this conjecture with data gathered on a glass/epoxy composite [41]. The fatigue limit ε_m for the unreinforced epoxy and the failure strain for fiber–matrix debonding ε_{db} are indicated in the figure.

The schematic of the matrix damage mechanisms is shown in Figure 9.17 where the subcritical microcracks and their coalescence into a critical macrocrack are illustrated. These conjectured mechanisms were studied in greater detail in [42] where glass/epoxy [45/–45/0]$_s$ laminates were loaded by cyclic longitudinal tension. This loading induced matrix-dominated damage in the 45° plies. By carefully observing the fatigue damage development in the outer 45° plies, the authors found that before the microcracks conjectured and depicted in Figure 9.17 occurred, smaller microcracks were initiated with their planes inclined to the fiber direction, as illustrated in Figure 9.19. These microcracks were found to lie normal to the local maximum principal stress (LMPS) given by the angle β_c shown in the figure. The local stress state in the matrix consists of the stress σ_2 normal to the fibers and the in-plane shear stress σ_6 as illustrated in the figure. It was found that the LMPS-governed microcrack initiation depended on the ratio $\lambda_{12} = \sigma_6/\sigma_2$. For low values of this ratio, i.e., when the shear stress is low, e.g., when the off-axis angle θ approaches 90°, the

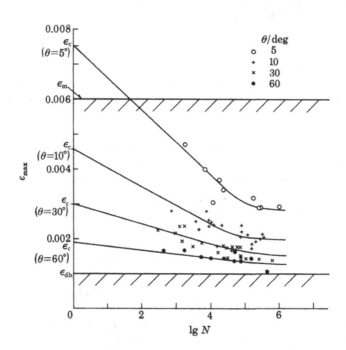

FIGURE 9.18 The conceptual FLDs in Figure 9.17 as supported by data in Hashin, Z. and Rotem, A., 1973. A fatigue failure criterion for fiber reinforced materials. *Journal of Composite Materials*, 7(4), 448–464.

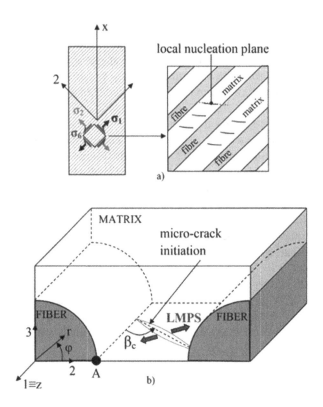

FIGURE 9.19 Microcracks nucleated along a plane oriented at angle β_c that lies normal to the local maximum principal stress (LMPS) [42]. Source: Quaresimin, M., Carraro, P.A. and Maragoni, L., 2016. Early stage damage in off-axis plies under fatigue loading. *Composites Science and Technology*, *128*, 147–154.

failure mechanism switches to fiber–matrix debonding. This agrees with the FLDs in Figure 9.18. This becomes clearer when the fatigue limit measured as ε_{max}, i.e., the maximum strain applied in the first load cycle, is plotted against the off-axis angle θ, as shown in Figure 9.20.

Two features are noteworthy in Figure 9.20. The first is that the fatigue limit of UD composites drops rapidly from θ = 0° until θ ≈ 60° beyond which it remains nearly constant, and the second is that the drop in fatigue limit of angle ply laminates also occurs rapidly until θ ≈ 60°. This suggests that the shear stress-induced damage is outcompeted by another mechanism operating beyond θ ≈ 60°. The authors of the study [42] identified that mechanism governed by the local hydrostatic stress (LHS) based on fatigue of [90]s UD composite tubes loaded in cyclic tension and torsion [43]. Their summary of data analyzed by the two criteria is shown in Figure 9.21 for 10^6 fatigue cycles. The separation of LMPS and LHS-governed failure initiation occurs at roughly θ = 60° in agreement with what is shown in Figure 9.20. As noted before, the fatigue limit shown in this figure is approximately the same for θ > 60° for both cases indicating that the governing mechanism is fiber–matrix debonding.

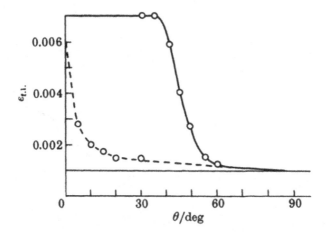

FIGURE 9.20 Fatigue limit $\varepsilon_{f.l.}$ plotted as a dashed line against the off-axis angle shown in Figure 9.17 based on the data shown there. Also shown is the fatigue limit of the same glass/epoxy angle ply $[\theta/-\theta]_s$ laminates [22]. Source: Talreja, R., 1981. Fatigue of composite materials: damage mechanisms and fatigue-life diagrams. *Proceedings of the Royal Society of London. A. Mathematical and Physical Sciences*, 378(1775), 461–475.

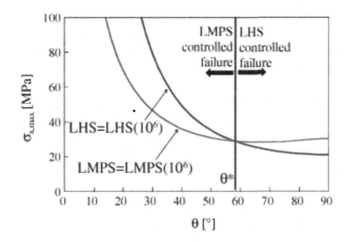

FIGURE 9.21 Constant life (10^6 cycles) diagram for off-axis loaded UD composite. Source: Quaresimin, M. and Carraro, P.A., 2014. Damage initiation and evolution in glass/epoxy tubes subjected to combined tension–torsion fatigue loading. *International Journal of Fatigue*, 63, 25–35.

As has been discussed in Chapter 8, Section 8.3, the local stress state in the matrix in transverse tension on UD composite is triaxial, and close to the fiber surface, it is hydrostatic tension. Thus, the critical failure mechanism here is brittle cavitation, governed by the critical dilatational energy density (or critical hydrostatic tension).

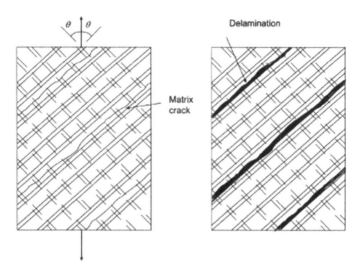

FIGURE 9.22   Fatigue damage in angle ply $[\theta/-\theta]_s$ laminates in cyclic axial tension. Source: Talreja, R., 1981. Fatigue of composite materials: damage mechanisms and fatigue-life diagrams. *Proceedings of the Royal Society of London. A. Mathematical and Physical Sciences, 378*(1775), 461–475.

As discussed in Chapter 8, it is reasonable to assume that the unstable growth of a cavity near the fiber surface will lead to fiber–matrix debonding.

Returning to Figure 9.20, the increased fatigue limit of angle ply $[\theta/-\theta]_s$ laminates in cyclic axial tension compared to the off-axis fatigue limit of UD composites is explained by two differences, as illustrated in Figure 9.22. First, the failure in angle ply laminates does not result from unstable growth of a single crack as in off-axis tension of UD composites due to the constraint effect of the adjacent plies, and secondly, the delamination mechanism in angle ply laminates is an additional energy dissipating mechanism to overcome before final failure.

9.2.3.1 Effect of Voids

Experimental studies of the effect of voids in initiating fatigue damage under combined loading in UD composites are made difficult by the inability to produce specimens with controlled voids and subjecting them to cyclic loads. It is more feasible to produce laminates with different void volume fractions and study the initiation of damage in the matrix of a selected ply within the laminate that is under combined loading. Such studies will be reviewed next.

9.3   LAMINATES

The angle ply laminates discussed above with respect to the fatigue limit compared to UD composites under off-axis cyclic loading bring out two aspects that need considering in case of fatigue of laminates: a) effect of ply constraint, and b) effect of ply/ply interfaces. These effects result in the damage development summarized in

Figure 9.2. The ply constraint effect results in multiple ply cracking that evolves with load cycles until a saturation state (CDS) is reached. This stage of damage development is reflected in longitudinal modulus reduction depicted in Figure 9.3. From CDS until the beginning of the dominance of fiber breakage, the damage is a complex phenomenon consisting of interconnected ply cracks and delamination. Once sufficient delamination has occurred, the separated plies undergo fatigue which is essentially what has been discussed in Section 9.2 on UD composites. In the following, we shall discuss the ply constraint effect on intralaminar cracking under cyclic loading and fatigue of ply/ply interfaces (i.e., interior delamination growth) following cracking within the plies.

The two effects involved in the fatigue damage development of composite laminates are best understood by considering multiple cracking of 90° plies within cross-ply laminates. In Chapter 7, Section 7.3.1, the observed mechanisms of damage in cross-ply laminates were described. Based on those, Figure 9.23 depicts three possible scenarios of the multiple transverse cracking under axial cyclic tension. The question to ask is: Where lies the mechanism of irreversibility that is essential for fatigue damage to evolve? In Scenario 1, the cracks are assumed with sharp tips at the 0/90 interfaces. Scenario 2 has cracks with delamination at the 0/90 interfaces. The delamination surfaces here undergo opening displacement and thus are traction free. Finally, Scenario 3 is like Scenario 2, except here the delamination surfaces are in contact, producing frictional shear stress on the surfaces as they conduct unequal axial displacement. Assuming elastic plies, Scenario 1 has no ability to provide damage evolution (crack density increase) because of the total reversibility under cyclic

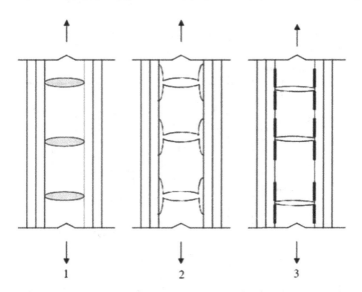

FIGURE 9.23   Three scenarios of transverse cracking in cross-ply laminates subjected to cyclic axial tension. Scenario 1: No delamination, Scenario 2: Delamination with traction-free surfaces, and Scenario 3: Delamination surfaces in frictional contact.

loading. In Scenario 2 where the delamination surfaces are traction free, it can be easily seen that it is equivalent to a reduced crack spacing (Figure 9.24). Thus, as the delamination length increases under fatigue, the maximum axial stress in the transverse plies midway between two adjacent cracks will decrease. Therefore, the driving force for producing new cracks will also reduce with cyclic loading, causing no increase in transverse crack density beyond what was produced in the first application of the axial load. In [44] Scenario 3 was analyzed using a variational approach to solve the unit cell boundary value problem with a specified (cubic) distribution of axial shear stress acting over the delamination surfaces. The delamination surfaces were grown with cyclic loading according to a Paris law for crack growth. The predicted crack density increase with cycles agreed with experimental data, as shown in Figure 9.25 [45].

In [45] a fatigue life prediction methodology for cross-ply laminates under axial tension was developed based on the prediction of transverse crack density evolution with cycles proposed in [44]. An empirical relationship was proposed for the crack density η_f reached just before the number of cycles to failure N_f, as

$$\eta_f = A \log N_f + B \tag{9.2}$$

where A and B are empirical constants.

In the progressive damage regime of fatigue failure signified by Region II of FLD of a cross-ply laminate described in Chapter 7, Section 7.3.1, it is assumed that failure in fatigue occurs when the crack density developed in fatigue reaches the critical crack density under static loading. The critical crack density is assumed to vary

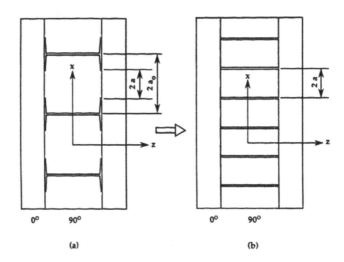

FIGURE 9.24 Scenario 2 of Figure 9.23 shown here with crack spacing $2a_0$ in (a) is equivalent to crack spacing $2a$ in (b). Source: Akshantala, N.V. and Talreja, R., 1998. A mechanistic model for fatigue damage evolution in composite laminates. *Mechanics of Materials*, 29(2), 123–140.

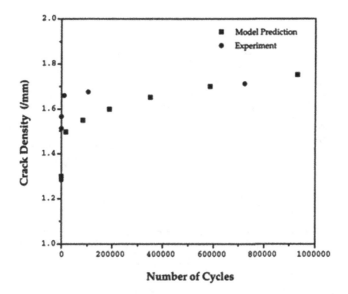

FIGURE 9.25 Transverse crack evolution with cycles for a [0/90₂]ₛ carbon/epoxy lami-nate at a maximum stress of 482.633 MPa and a stress ratio of 0.1. Experimental data from Charewicz, A. and I.M. Daniel, I.M. 1986. Damage mechanisms and accumulation in graph-ite/epoxy laminates. H.T. Hahn (Ed.), Composite Materials: Fatigue and Fracture, ASTM STP 907, 2, American Society for Testing and Materials, Philadelphia, 274–297.

between its maximum value η_c obtained at the upper end of Region II of the FLD, given by the static strain to failure ε_c, and its minimum value η_{fpf} is assumed to cor-respond to the fatigue limit given by the first ply failure strain ε_{fpf}. As an illustration for a glass/epoxy cross-ply laminate, the number of cycles to failure at the beginning of Region II is taken as $N_f = 10^2$ cycles, and the fatigue limit is placed at $N_f = 10^6$ cycles, as shown in Figure 9.26. The constants A and B appearing in Equation (9.2) are obtained by setting the crack densities η_c and η_{fpf} in Equation (9.2) at $N_f = 10^2$ and $N_f = 10^6$ cycles, respectively. Thus,

$$\eta_c = 2A + B \tag{9.3}$$

$$\eta_{fpf} = 6A + B \tag{9.4}$$

Solving Equations (9.3) and (9.44) provides values of A and B in terms of the mea-sured crack densities.

Using the data reported in [47], the constants A and B were determined, and the procedure depicted in Figure 9.26 was completed as shown in Figure 9.27. Finally, the FLD was generated as shown in Figure 9.28 with fatigue life data inserted from [46].

In any fatigue damage modeling, it is important at the outset to be clear on where the irreversibility in cyclic loading is assumed. In the modeling just described, the irreversible (energy dissipating) mechanism is assumed to be in friction between the

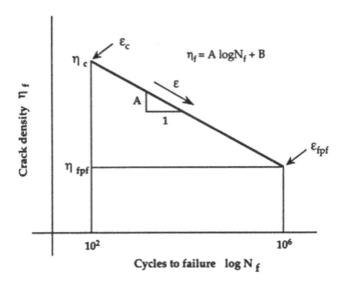

FIGURE 9.26 A schematic depiction of the failure criterion in terms of the transverse crack density for a glass/epoxy cross-ply laminate. Source: Akshantala, N.V. and Talreja, R., 2000. A micromechanics based model for predicting fatigue life of composite laminates. *Materials Science and Engineering: A*, 285(1–2), 303–313.

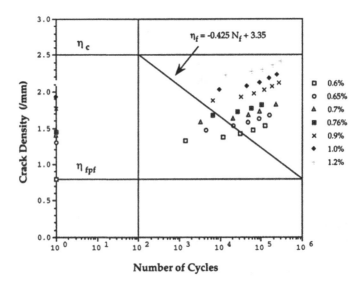

FIGURE 9.27 Semi-log plot of change in crack density with cycles for a [0/90]$_s$ glass/epoxy laminate at different maximum strain levels. Source: Akshantala, N.V. and Talreja, R., 2000. A micromechanics based model for predicting fatigue life of composite laminates. *Materials Science and Engineering: A*, 285(1–2), 303–313.

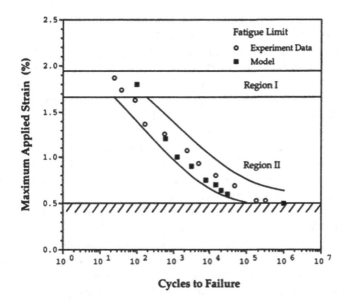

FIGURE 9.28   Fatigue life diagram for a [0/90]$_s$ glass/epoxy laminate [45]. Experimental fatigue life data from [48]. Source: Akshantala, N.V. and Talreja, R., 2000. A microme-chanics based model for predicting fatigue life of composite laminates. *Materials Science and Engineering: A, 285*(1–2), 303–313, with experimental data from Charewicz, A. and I.M. Daniel, I.M. 1986. Damage mechanisms and accumulation in graphite/epoxy laminates. H.T. Hahn (Ed.), Composite Materials: Fatigue and Fracture, ASTM STP 907, 2, American Society for Testing and Materials, Philadelphia, 274–297.

two faces of the delamination at the 0/90 interfaces and by extension between two delaminated surfaces of differently oriented plies. The geometrical model set up assumed all ply cracks to have been fully grown across the ply thickness and to have caused delamination at their fronts as shown in Scenarios 2 and 3 in Figure 9.23. Scenario 1 in that figure was argued to be incapable of causing crack density increase with load cycles because of the absence of irreversibility with assumed elasticity. A work [48] assumed the cracks of Scenario 1 but assumed the cracks not to be fully grown in the width direction when initiated. The irreversibility with load cycles was attributed to the growth of these cracks along the width, i.e., the fiber direction in a ply. For this growth, a power law like the common Paris law for crack growth rate was assumed. The energy release rates for the generally mixed-mode crack growth were calculated accounting for the mutual interaction of the ply cracks. To account for cracks of different lengths growing in the fiber direction a crack density measure was defined as the ratio of the added-up length of all cracks divided by the surface area of the specimen over which cracks were measured.

In the later work [49] the authors of [48] included delamination at the fronts of the ply cracks in their fatigue damage evolution analysis. The delamination growth was analyzed with linear elastic fracture mechanics for cross-ply laminates assuming equal growth of delamination length as idealized in Scenarios 2 and 3 in Figure 9.23.

Numerical calculation of energy release rates (ERRs) by the virtual crack closure technique (VCCT) indicated that the mode II component of the mixed-mode ERR at the delamination front was dominant, supporting Scenario 3. Figure 9.29 shows example of the variation of the mode II ERR normalized by the square of the applied axial stress with the delamination ratio (defined as the area of delamination divided by the area of observation on the specimen surface) for different transverse crack densities. As seen in the figure, the driving force for delamination growth takes a maximum before declining to zero. The decrease in the ERR is attributed to the shielding effect of the delamination from an adjacent ply crack.

The results depicted in Figure 9.29 can be inferred in terms of the effect of delamination growth on the further evolution (multiplication) of transverse cracks. As shown, as the crack density increases from $\rho = 0.2$ to 0.8 cracks per unit length, i.e., as the cracks get closer, the maximum mode II ERR reduces. Thus, the reduced crack driving force slows down the delamination growth. Then, as the increase per load cycle in the delamination length l reduces, the rate at which a new crack initiates between two preexisting cracks also slows down. This process is illustrated in Figures 9.30 and 9.31 [50]. As schematically shown in Figure 9.30, a new crack between two preexisting cracks forms midway between the cracks induced by the axial normal stress, which is maximum at that point. This stress changes when the delamination length $2l$ increases. Figure 9.31 compares the change in this stress with an increase in the delamination length. The figure to the left is for the case of traction-free delamination surfaces and the one to the right is for a constant shear stress on the delamination surfaces. As seen in the figure, the axial stress will not produce new cracks if the delamination surfaces are traction free, while the presence of shear stress on the delamination surfaces increases the axial normal stress, thereby allowing new cracks to form midway between two pre-existing transverse cracks. In this

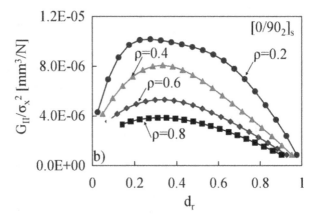

FIGURE 9.29 Variation of normalized mode II ERR for a $[0/90_2]_s$ glass/epoxy laminate with the normalized delamination area ratio d_r for different crack densities ρ in cracks/mm. Source: Carraro, P.A., Maragoni, L. and Quaresimin, M., 2017. Prediction of the crack density evolution in multidirectional laminates under fatigue loadings. *Composites Science and Technology*, *145*, 24–39.

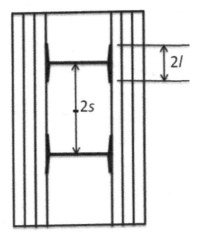

FIGURE 9.30 The formation of a new crack between two preexisting transverse cracks at spacing $2s$ is induced by the maximum axial normal stress midway between the cracks when the delamination length $2l$ increases.

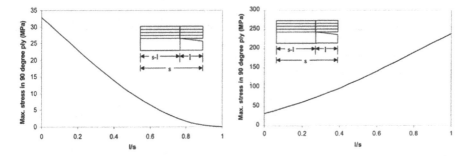

FIGURE 9.31 Variation of the maximum axial normal stress in the 90° ply midway between two preexisting transverse cracks of spacing $2s$ (Figure 9.30) as the delamination length l increases: (left) traction-free delamination surfaces, (right) constant shear stress on delamination surfaces. Source: Talreja, R., 2006. Damage analysis for structural integrity and durability of composite materials. *Fatigue & Fracture of Engineering Materials & Structures*, 29(7), 481–506.

argument, the shear stress is assumed constant for illustration but in the model presented in [44], it was assumed to vary with a cubic variation. The resulting crack density evolution is shown in Figure 9.32, which qualitatively agrees with experimental data in Figure 9.33 reported by [51].

9.3.1 Effects of Manufacturing Defects

Phenomenological models aimed at providing knockdown factors for the effect of manufacturing defects on fatigue life are of limited use in view of the variety of

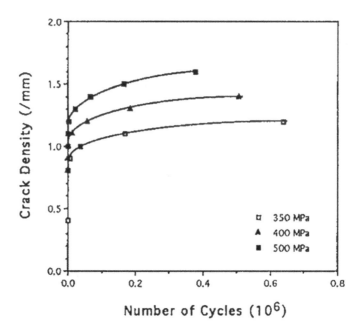

FIGURE 9.32 Crack density prediction at different fatigue load levels. Source: Akshantala, N.V. and Talreja, R., 2000. A micromechanics based model for predicting fatigue life of composite laminates. *Materials Science and Engineering: A*, 285(1-2), pp.303-313.

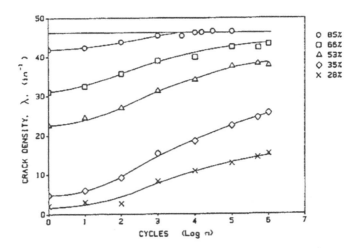

FIGURE 9.33 Experimental for crack density in cross-ply laminates at different fatigue load levels given as percentage of the ultimate tensile load. Source: Diao, X., Ye, L. and Mai, Y. -W., 1995. Simulation of fatigue performance of cross-ply laminates. *Appl. Compos. Mater.* 3, 391– 406.

effects found experimentally depending on the manufacturing process, fiber architecture, matrix polymer, etc. As reviewed in Chapter 7, Section 7.4.2, cross-ply laminates are suitable for studying the effects of voids on the early stages of fatigue damage within the plies, transverse crack formation and multiplication, as well as the local delamination caused by the cracks at ply interfaces. The early stages of fatigue damage are affected similarly in plies within a laminate as in a UD composite, but they can be studied in more detail under the constraint of adjacent plies that helps slow down the damage process. At the same time, computational micromechanics can aid in understanding the effects on the local stress fields by the presence of voids in different sizes, shapes, and distributions.

9.3.1.1 Ply Cracking

The experimental observations reported in [52] indicated that the presence of voids affected the multiple cracking processes in cross-ply laminates and in off-axis (45°) layers of laminates prior to achieving saturation crack densities but did not significantly affect the saturation crack density states. In [48] the authors used a crack density measure where partial cracks were accounted for, and this measure showed a significant effect of the presence of voids within the plies. The experimental data for transverse cracking in cross-ply laminates were displayed in Figure 7.22 (Chapter 7) where the effect of voids is seen to begin the cracking process earlier and at higher rates than in void-free specimens.

As has been discussed in the context of static tension normal to fibers (Figures 8.43 and 8.44, Chapter 8), voids in the matrix initiate brittle cavitation, and therefore fiber–matrix debonding, earlier by inducing more favorable conditions for dilatational energy density near fiber surfaces. In a study [53] elongated voids of spherical and elliptical cross sections were placed at different locations within an RVE of nonuniform fiber distribution, as shown in Figure 9.34. It was found that the voids generally shifted the position of brittle cavitation in the matrix, still close to fiber surfaces, as compared to when no voids were present, as illustrated in the figure. Also, when an elliptical void was made slim (of a high aspect ratio), it affected the local stress state such that the condition for brittle cavitation was no longer favorable. Instead, yielding occurred at the tips of the elliptical void.

In a study [54] where an elongated (cigar-shaped) void was placed between fibers in an off-axis (45°) ply, it was found that the initiation of microcracks between fibers was as it would be without the void, i.e., in planes lying normal to the local maximum principal stress. However, in actual specimens with distributed voids, it was found that the number of initiated microcracks under fatigue was much higher compared to when the specimens were void free. To study the initiation of the first macrocrack formed by the coalescence of such microcracks two different laminate stacking sequences, namely $[0/90_2]_S$ and $[0/45_2/0/-45_2]_S$, were considered [55]. A computational FE modeling was performed to determine the maximum values of the local hydrostatic stress (equivalent to the dilatational energy density) and the local principal stress in RVEs with distributed voids. For the case of the cross-ply laminate under cyclic axial tension, the number of cycles to initiation of a macrocrack

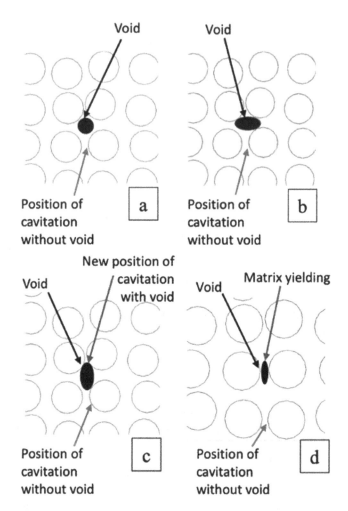

FIGURE 9.34 Voids are placed in different locations within the matrix of RVEs. (a) circular void, (b) an elliptical void with a major axis aligned with the loading direction, (c) an elliptical void with a major axis normal to the loading direction, and (d) a slim elliptical void with a major axis normal to the loading direction. Source: Elnekhaily, S.E., 2018. Damage initiation in unidirectional polymeric composites with manufacturing induced irregularities, PhD dissertation, Department of Materials Science and Engineering, Texas A&M University, College Station, Texas.

formed by coalescence of fiber–matrix debonding was found to correlate with maximum local hydrostatic stress (LHS*) calculated as averages over several RVEs in a control volume. Similarly, for the life to initiation of a macrocrack along fibers by coalescence of inclined microcracks in −45 ply of the $[0/45_2/0/−45_2]_S$ laminate, the correlation with the RVE-averaged local maximum principal stress (LMPS*) was found. Both cases are shown in Figure 9.35 where the collapsing data for voids and

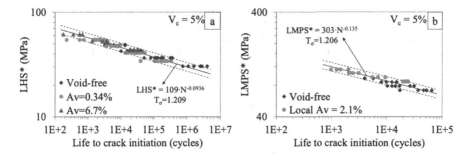

FIGURE 9.35 Life to crack initiation in terms of the a) RVE-averaged local hydrostatic stress (LHS*) in 90° ply of [0/90₂]ₛ laminates, b) RVE-averaged local maximum principal stress (LMPS*) in –45° ply of [0/45₂/0/–45₂]ₛ laminate. Source: Maragoni, L., Carraro, P.A. and Quaresimin, M., 2019. Prediction of fatigue life to crack initiation in unidirectional plies containing voids. *Composites Part A: Applied Science and Manufacturing*, *127*, 105638.

void-free laminates are found to collapse in scatter bands while when plotted against the applied stress the data separate in distinct scatter bands.

9.4   CONCLUSION

In Chapter 7, the mechanisms of damage occurring in UD composites and in laminates under cyclic loading were reviewed. The observations and measurements made experimentally were interpreted using a systematic conceptual framework called fatigue life diagrams (FLDs) developed by this author. To focus on the effect of manufacturing defects, that framework was used to describe changes caused by defects in the baseline behavior. In the current chapter, various modeling strategies developed to predict fatigue damage evolution and its criticality (lifetime) have been discussed with particular attention to the effects of manufacturing defects.

REFERENCES

1. Jamison, R.D., Schulte, K., Reifsnider, K.L. and Stinchcomb, W.W., 1984. Characterization and analysis of damage mechanisms in tension-tension fatigue of graphite/epoxy laminates. *Effects of Defects in Composite Materials, ASTM STP*, *836*, 21–55.
2. Talreja, R. 1985. A Continuum mechanics characterization of damage in composite materials. *Proceedings of the Royal Society of London*, *A399*, pp. 195–216.
3. Talreja, R. 1985. Transverse cracking and stiffness reduction in composite laminates. *Journal of Computational and Applied Mathematics*, *19*, pp. 355–375.
4. Talreja, R. 1986. Stiffness properties of composite laminates with matrix cracking and interior delamination. *Engineering Fracture Mechanics*, *25*(5/6), 751–762.
5. Talreja, R. 1989. Damage development in composites: Mechanisms and modelling. *Journal of Strain Analysis*, *4*(4), pp. 215–222.
6. Talreja, R. 1989. Continuum modelling of the development of intralaminar cracking in composite laminates. In Salama, K. et al. (eds.), *Advances in Fracture Research*, Vol. 3. Pergamon Press, Oxford, pp. 2191–2199.

7. Talreja, R. 1990. Internal variable damage mechanics of composite materials. In Boehler, J.P. (ed.), *Invited Paper Yielding, Damage and Failure of Anisotropic Solids.* Mechanical Engineering Publications, London, pp. 509–533.

8. Ladeveze, P., 1993, Inelastic strain and damage modelling. In Talreja, R. (ed.), *Damage Mechanics of Composites Materials.* Amsterdam: Elsevier.

9. Matzenmiller, A.L.J.T.R., Lubliner, J. and Taylor, R.L., 1995. A constitutive model for anisotropic damage in fiber-composites. *Mechanics of materials*, 20(2), pp. 125–152.

10. Maire, J.F. and Chaboche, J.L., 1997. A new formulation of continuum damage mechanics (CDM) for composite materials. *Aerospace Science and Technology*, 1(4), pp. 247–257.

11. Maimí, P., Camanho, P.P., Mayugo, J.A. and Dávila, C.G., 2007. A continuum damage model for composite laminates: Part I–Constitutive model. *Mechanics of Materials*, 39(10), pp. 897–908.

12. Varna, J., Joffe, R., Akshantala, N.V. and Talreja, R., 1999. Damage in composite laminates with off-axis plies. *Composites Science and Technology*, 59(14), pp. 2139–2147.

13. Varna, J., Joffe, R. and Talreja, R., 2001. A synergistic damage-mechanics analysis of transverse cracking in [±θ/904] s laminates. *Composites Science and Technology*, 61(5), pp. 657–665.

14. Varna, J., Joffe, R. and Talreja, R., 2001. Mixed micromechanics and continuum damage mechanics approach to transverse cracking in [S, 90 n] s laminates. *Mechanics of Composite Materials*, 37, pp. 115–126.

15. Varna, J., Krasnikovs, A., Kumar, R.S. and Talreja, R., 2004. A synergistic damage mechanics approach to viscoelastic response of cracked cross-ply laminates. *International Journal of Damage Mechanics*, 13(4), pp. 301–334.

16. Varna, J. and Talreja, R., 2012. Integration of macro-and microdamage mechanics for the performance evaluation of composite materials. *Mechanics of Composite Materials*, 48, pp. 145–160.

17. Varna, J., 2008. Physical interpretation of parameters in synergistic continuum damage mechanics model for laminates. *Composites Science and Technology*, 68(13), pp. 2592–2600.

18. Reifsnider, K.L., 1986. The critical element model: A modeling philosophy. *Engineering Fracture Mechanics*, 25(5–6), pp. 739–749.

19. Gamstedt, E.K. and Talreja, R., 1999. Fatigue damage mechanisms in unidirectional carbon-fibre-reinforced plastics. *Journal of Materials Science*, 34, pp. 2535–2546.

20. Castro, O., Carraro, P.A., Maragoni, L. and Quaresimin, M., 2019. Fatigue damage evolution in unidirectional glass/epoxy composites under a cyclic load. *Polymer Testing*, 74: 216–224.

21. Dharan, C.K.H., 1975. Fatigue failure in graphite fibre and glass fibre-polymer composites. *Journal of Materials Science*, 10, 1665–1670.

22. Talreja, R., 1981. Fatigue of composite materials: Damage mechanisms and fatigue-life diagrams. *Proceedings of the Royal Society of London. A. Mathematical and Physical Sciences*, 378(1775), pp. 461–475.

23. Aroush, D.R.B., Maire, E., Gauthier, C., Youssef, S., Cloetens, P. and Wagner, H.D., 2006. A study of fracture of unidirectional composites using in situ high-resolution synchrotron X-ray microtomography. *Composites Science and Technology*, 66(10), pp. 1348–1353.

24. Scott, A.E., Mavrogordato, M., Wright, P., Sinclair, I. and Spearing, S.M., 2011. In situ fibre fracture measurement in carbon–epoxy laminates using high resolution computed tomography. *Composites Science and Technology*, 71(12), pp. 1471–1477.

25. Zhuang, L., Talreja, R. and Varna, J., 2016. Tensile failure of unidirectional composites from a local fracture plane. *Composites Science and Technology*, 133, pp. 119–127.

26. Pagano, F., 2019. *Mécanismes de fatigue dominés par les fibres dans les composites stratifiés d'unidirectionnels* (Doctoral dissertation, Paris Sciences et Lettres (ComUE)).
27. Alves, M. and Pimenta, S., 2018. A computationally-efficient micromechanical model for the fatigue life of unidirectional composites under tension-tension loading. *International Journal of Fatigue, 116*, pp. 677–690.
28. Pimenta, S., Mersch, A. and Alves, M., 2018, July. Predicting damage accumulation and fatigue life of UD composites under longitudinal tension. In *IOP Conference Series: Materials Science and Engineering* (Vol. 388, No. 1, 012007). IOP Publishing.
29. Pimenta, S., Pinho, S.T., 2013. Hierarchical scaling law for the strength of composite fibre bundles. *Journal of the Mechanics and Physics of Solids, 61*(6), pp. 1337–1356.
30. Sørensen,B.F., S.Goutianos. 2019. Micromechanical model for prediction of the fatigue limit for unidirectional fibre composites. *Mechanics of Materials, 131*, pp. 169–187.
31. Korkiakoski, S., Brøndsted, P., Sarlin, E. and Saarela, O., 2016. Influence of specimen type and reinforcement on measured tension–tension fatigue life of unidirectional GFRP laminates. *International Journal of Fatigue, 85*, pp. 114–129.
32. Pisanova, E.V., Zhandarov, S.F. and Dovgyalo, V.A., 1994. Interfacial adhesion and failure modes in single filament thermoplastic composites. *Polymer Composites, 15*(2), pp. 147–155.
33. Ten Busschen, A. and Selvadurai, A.P.S., 1995. Mechanics of the segmentation of an embedded fiber, part I: experimental investigations. *Journal of Applied Mechanics, Transactions ASME, 62*, pp. 87–97.
34. Huang, H. and Talreja, R., 2006. Numerical simulation of matrix micro-cracking in short fiber reinforced polymer composites: Initiation and propagation. *Composites Science and Technology, 66*(15), pp. 2743–2757.
35. Rosen, B.W., 1964. *Mechanics of Composite Strengthening: Fiber Composites.* Materials Park, OH: American Society of Metals, pp. 37–45.
36. Dadkhah, M.S., Cox, B.N. and Morris, W.L., 1995. Compression-compression fatigue of 3D woven composites. *Acta Metallurgica et Materialia, 43*(12), pp. 4235–4245.
37. Argon, A.S. 1972. Fracture of composites. In H. Herman (ed.),*Treatise of Materials Science and Technology*, Vol. 1. Academic Pres, New York, pp. 79–114.
38. Liebig, W.V., Viets, C., Schulte, K. and Fiedler, B., 2015. Influence of voids on the compressive failure behaviour of fibre-reinforced composites. *Composites Science and Technology, 117*, pp. 225–233.
39. Quaresimin, M., 2015. 50th anniversary article: Multiaxial fatigue testing of composites: From the pioneers to future directions. *Strain, 51*(1), pp. 16–29.
40. Quaresimin, M., Susmel, L. and Talreja, R., 2010. Fatigue behaviour and life assessment of composite laminates under multiaxial loadings. *International Journal of Fatigue, 32*, pp. 2–16.
41. Hashin, Z. and Rotem, A., 1973. A fatigue failure criterion for fiber reinforced materials. *Journal of Composite Materials, 7*(4), pp. 448–464.
42. Quaresimin, M., Carraro, P.A. and Maragoni, L., 2016. Early stage damage in off-axis plies under fatigue loading. *Composites Science and Technology, 128*, pp. 147–154.
43. Quaresimin, M. and Carraro, P.A., 2014. Damage initiation and evolution in glass/epoxy tubes subjected to combined tension–torsion fatigue loading. *International Journal of Fatigue, 63*, pp. 25–35.
44. Akshantala, N.V. and Talreja, R., 1998. A mechanistic model for fatigue damage evolution in composite laminates. *Mechanics of Materials, 29*(2), pp. 123–140.
45. Akshantala, N.V. and Talreja, R., 2000. A micromechanics based model for predicting fatigue life of composite laminates. *Materials Science and Engineering: A, 285*(1–2), pp. 303–313.

46. Charewicz, A. and I.M. Daniel 1986. Damage mechanisms and accumulation in graph-ite/epoxy laminates. In Hahn, H.T. (Ed.), *Composite Materials: Fatigue and Fracture*, ASTM STP 907, 2, American Society for Testing and Materials, Philadelphia, pp. 274–297.
47. Varna, J. and Berglund, L.A., 1994. Thermo-elastic properties of composite laminates with transverse cracks. *Journal of Composites Technology and Research*, 16, 77–87.
48. Carraro, P.A., Maragoni, L. and Quaresimin, M., 2017. Prediction of the crack density evolution in multidirectional laminates under fatigue loadings. *Composites Science and Technology*, *145*, pp. 24–39.
49. Carraro, P.A., Maragoni, L. and Quaresimin, M., 2019. Characterisation and analysis of transverse crack-induced delamination in cross-ply composite laminates under fatigue loadings. *International Journal of Fatigue*, 129, 105217.
50. Talreja, R., 2006. Damage analysis for structural integrity and durability of compos-ite materials. *Fatigue & Fracture of Engineering Materials & Structures*, 29(7), pp. 481–506.
51. Diao, X., Ye, L. and Mai, Y. -W., 1995. Simulation of fatigue performance of cross-ply laminates. *Applied Composite Materials*, *3*, 391–406.
52. Sisodia, S.M., Gamstedt, E.K., Edgren, F. and Varna, J., 2015. Effects of voids on quasi-static and tension fatigue behaviour of carbon-fibre composite laminates. *Journal of Composite Materials*, *49*, pp. 2137–2148.
53. Elnekhaily, S.E., 2018. Damage initiation in unidirectional polymeric composites with manufacturing induced irregularities, PhD dissertation, Department of Materials Science and Engineering, Texas A&M University, College Station, Texas.
54. Maragoni, L., Carraro, P.A. and Quaresimin, M., 2016. Effect of voids on the crack formation in a [45/−45/0] s laminate under cyclic axial tension. *Composites Part A: Applied Science and Manufacturing*, *91*, pp. 493–500.
55. Maragoni, L., Carraro, P.A. and Quaresimin, M., 2019. Prediction of fatigue life to crack initiation in unidirectional plies containing voids. *Composites Part A: Applied Science and Manufacturing*, *127*, 105638.

Index

Printed in the United States
by Baker & Taylor Publisher Services